Math for Chemistry

A Comprehensive Guide to the Mathematical Concepts in College level Chemistry courses

WRITTEN AND PUBLISHED BY: Maria Diss Rosero, PhD

EDITED BY: Andrew Delaney and Sarah Warren

ISBN: 979-8-9991406-0-9

Preface

The purpose of this book is to provide a mathematics reference guide that is easily accessible for students taking chemistry. Foundational information required for solving chemistry problems is present to facilitate recovery or development of necessary math skills. This book is designed to help with the math involved in high school and college level chemistry, general chemistry and nursing chemistry courses.

Each section includes a brief explanation and practice problems. I have found that practice is essential to understand concepts required in solving math and science problems. Additionally, as your studies become increasingly specialized, there will be few problems or problem types on which you can be tested. Consequently, there is a high probability that problems similar to the practice problems in this book will appear on your quizzes, tests and exams.

Table of Contents

General Review

Addition and Subtraction

Basic math operations, you will need to know adding, subtracting, multiplying, and dividing. These are the foundational operations for all the math problems you will encounter in your chemistry courses. More than likely, you will not need to be great at doing these basic operations in your head as you will likely be allowed to use a calculator, but you should at least know how to do these. (1.1)

$$17 + 21 = \qquad\qquad 5 * 3 =$$

$$29 - 13 = \qquad\qquad 9 \div 3 =$$

$$8 \times 7 = \qquad\qquad \frac{32}{8} =$$

$$\frac{21}{7} = \qquad\qquad (32)(4) =$$

Multiplication can be denoted in a couple of ways as shown in the previous problems. Whether the symbol is ×, *, or ( )( ), the order of multiplication is implied. Similarly, division can be denoted with a ÷, / or with a fraction. The problem will indicate whether a decimal or a simplified fraction of the same value should be inputted as the answer. We will discuss decimal places in the next section.

Negative numbers may come up in calculations, recall the number line (Figure 1) negative numbers are less than positive ones and the same distance from zero as positive ones, just in the opposite direction.

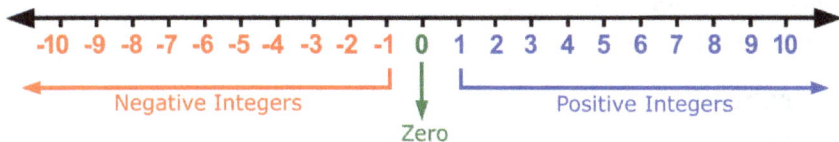

Figure 1. Number line with negative and positive integers equidistant from zero.

Here are a few sample questions for you to practice (1.2).

$$4 + -7 = \qquad\qquad -6 \div -2 =$$

$$-5 + -5 = \qquad\qquad 5 \times -6 =$$

$$31 - (-27) = \qquad\qquad -7 \times (-3) =$$

Adding one negative number to a positive number is essentially subtracting the negative number from the positive one, as the negative number is less than the positive one.

Adding two negative numbers together makes a number that is more negative.

Two negative signs cancel out, so subtracting a negative number is the same as adding the positive number.

When multiplying or dividing two negative numbers give a positive answer while one negative number gives a negative answer.

Order of Operations

Next, let's work on one of the trickiest things about solving problems for beginners which is knowing the order of operations. There is a specific order of operations to solving these problems: first do any math operations in parentheses, then do exponents, then multiply, then divide, then add and finally subtract. It is important to note that while calculators are excellent tools for computation, keystroke sequence is important so knowing what order of operations to do for the problem is necessary. A handy acronym for this is PEMDAS, I learned the mnemonic "Please excuse my dear Aunt Sally" to help me remember this.

Parentheses $\rightarrow$ **Exponents** $\rightarrow$ **Multiplication** $\rightarrow$ **Division** $\rightarrow$ **Addition** $\rightarrow$ **Subtraction**

The order of operations is the most important part of solving problems. Doing the incorrect order of operations often leads to the wrong answer.

Consider:

$$6 + 7 \times 3 =$$

The operations that need to be carried out are **Addition** and **Multiplication**. If one follows PEMDAS, the first operation you do is **Multiplication**: multiply 7 by 3 and get 21. Then you would do **Addition**: add 6 to 21, which gives you 27. If you add 6 plus 7 and then multiply by 3, you get 39 which is not the correct answer as it did not follow the correct order of operations.

$$6 + 7 \times 3 = 27$$

Consider:

$$9 \times 3 + 7 - 10 \div 5 =$$

Looking at the above equation, we have 4 different mathematical operations: **Multiplication**, **Addition**, **Subtraction** and **Division**. Using the correct order of operations (PEMDAS), we need to perform the operations as follows: **Multiplication**, **Division**, **Addition** then **Subtraction**.

$$(9 \times 3) + 7 - 10 \div 5 =$$

Our first step is to do the **Multiplication**, so we can group the multiplication in parentheses to see the operation more clearly. Once we multiply 9 x 3 we get 27, so we can replace the above parentheses with 27. Now we can move on to the **Division** operation.

$$27 + 7 - (10 \div 5) =$$

Once again, we use parentheses to group the division operation to see it more clearly. Ten divided by five is 2, so we can replace the above parenthesis with 2.

$$27 + 7 - 2 =$$

The remaining operations are **Additional** and **Subtraction**. 27 + 7 gives us 34.

$$34 - 2 =$$

By subtracting 2 from 34 we get our final answer of 32.

Here are a few general practice problems (1.3)

$$(11 + -9) \times 4 + 5 - 21 \div 7 =$$

$$(37 - 25) \div 6 + 7 =$$

$$3^2 - 16 \div 4 + 32 =$$

Here are a few chemistry related practice problems

$$K = {}^0C + 273.15$$

To convert degrees Celsius to Kelvin, we use the above equation. To practice let's covert the following:

$$154^0C \text{ to } K$$
$$-60^0C \text{ to } K$$

To solve these, we plug in the value given for ^0C into the $K = {}^0C + 273.15$ equation to obtain the temperature in Kelvin.

$$K = 154 + 273.15 = 427.15$$
$$K = -60 + 273.15 = 213.15$$

Another type of temperature conversion you will encounter will include converting degrees Fahrenheit to degrees Celsius. The conversion equation for that is shown below:

$$(\mathbf{{}^{\circ}F} - 32) \times 5/9 = \mathbf{{}^{\circ}C}$$

Given the temperature conversion formula above, convert the following:

$$176^0F \text{ to } {}^0C$$

$$-6^0F \text{ to } {}^0C$$

To solve for these temperature conversions, we would plug the value of ^0F into the equation:

$${}^0C = ({}^0F - 32) \times 5/9$$
$${}^0C = (176 - 32) \times 5/9 = 80 \ {}^0C$$
$${}^0C = (-6 - 32) \times 5/9 = -21 \ {}^0C$$

Now if you were asked to covert K to ^0C, or ^0C to ^0F, you would need to rearrange the corresponding equations to solve for the desired temperature either ^0C or ^0F, let's review how to do that.

Operating on Both Sides of an Equation and Isolating Variables

The formulas you will come across in chemistry are usually set up with numbers and variables on both sides of the equal sign as we have seen in the above examples. To solve for one variable, you will need to isolate it on one side of the equation. Since both sides of the equation are equal, we need to do the same thing on both sides to hold the same value.

Let's consider the Kelvin to Celsius equation as an example, $K = {}^0C + 273.15$. If we want to solve for Celsius, we will need to get it on one side of the equation by itself. To accomplish this, we will need to do the opposite operation on both sides of the equation.

To get "C" by itself we will need to get rid of the "+ 273.15." To do this, we will subtract 273.15 from both sides of the equation which will leave us with:

$$K - 273.15 = {}^0C + 273.15 - 273.15$$

$$K - 273.15 = {}^0C$$

Which can be rewritten as $^0C = K - 273.15$

With the equation rewritten to solve for degrees Celsius we are now able to convert K to ^0C.

So, what is the temperature in Celsius if the temperature in Kelvin is 300 K?

We would use the equation $^0C = K - 273.15$ and plug in 300 for K to get:

$^0C = 300 - 273.15 = 26.85$

Let's consider the conversion of Celsius to Fahrenheit. To solve for ^0F, we will need to isolate ^0F on one side of the equation and whatever is on the other side of the equation are the operations

9

necessary to solve for °F. Since both sides of the equation are equal, we need to do the same thing on both sides to hold the same value.

$$(°F - 32) \times \frac{5}{9} = °C$$

This equation can be re-written as

$$\frac{5 \times (°F - 32)}{9} = °C$$

To isolate °F, we first need to multiply each side of the equation by 9

$$\frac{5 \times (°F - 32)}{9} \times 9 = 9 \times °C$$

Which can be simplified to

$$5 \times (°F - 32) = 9 \times °C$$

Next, we need to divide by sides of the equation by 5,

$$\frac{5 \times (°F - 32)}{5} = \frac{9 \times °C}{5}$$

Which simplifies to

$$°F - 32 = \frac{9 \times °C}{5}$$

Next, to isolate °F, we need to add 32 to each side of the equation

$$°F - 32 + 32 = \frac{9 \times °C}{5} + 32$$

Which can be simplified to

$$°F = \frac{9 \times °C}{5} + 32$$

With the equation rearranged and set up to solve for degrees Fahrenheit we see that we can now easily covert degrees Celsius to degrees Fahrenheit.

What is the temperature in Fahrenheit given the temperature in Celsius is 100 degrees?

To solve for this, we would plug the value of Celsius into the following equation

$$°F = \frac{9 \times °C}{5} + 32 = \frac{9 \times 100}{5} + 32$$
$$°F = 212$$

So, when the temperature is 100 °C, it is also 212 °F.

There are countless equations in chemistry that have more than one variable and require you to solve for other variables. Let's do a few more examples.

The ideal gas law is PV = nRT, where P is pressure (in atm), V is volume (in liters), n is number of moles and T is temperature (in K). R is a constant and is equal to 0.08206 $\frac{L \times atm}{mol \times K}$.

Let's solve the equation in terms of T (temperature).

To do this we will need to isolate T on one side.

PV = nRT

T is being multiplied by n and R. To cancel out these operations, we will need to divide both sides by n and R.

$$\frac{PV}{nR} = \frac{nRT}{nR}$$

The n and R on the side with the T cancel out so we are no left with

$$\frac{PV}{nR} = T$$

Which can be rewritten as

$$T = \frac{PV}{nR}$$

For additional practice, solve the ideal gas law for P, V and n. (1.4)

Additional rearranging equation problems (1.5)

Consider the dilution equation: $M_1V_1 = M_2V_2$, were M_1 and M_2 are molar concentrations and V_1 and V_2 are volumes. Rearrange the equation to solve for V_2.

Consider Gibbs free energy equation: $\Delta G = \Delta H - T\Delta S$, were ΔG is the change in free energy, ΔH is the change in enthalpy, T is the temperature, and ΔS is the change in entropy. Rearrange the equation and solve for ΔS, ΔH and T.

Significant Digits

In chemistry, all measurements require some sort of instrument. The exactness of the measurement depends on the sensitivity of said instrument. To account for the sensitivity of the instrument and the uncertainty in the measurement, the digits we record are referred to as significant digits. The greater the number of significant digits in a measurement, the greater is the certainty of that measurement. For example, there are four significant digits in 2.058 kg and two in 2.0 kg.

To determine the number of significant digits, observe the following rules:

1. Make sure that the measured quantity has a decimal point.
2. Count the numbers going from the left to the right until you reach the first nonzero digit.
3. Count that digit and every digit to its right as significant.

When zeros are at the end of a measurement it can be a little tricky; zeros that end a number and lie either after or before the decimal point are significant. For instance, 1.070 L has four significant digits, and 1700. mL has four significant digits. When no decimal point is present such as with 1700 mL, it would be assumed that the zeros are not significant and exponential notation is needed to show the digits of significance. Therefore, 1700 mL would be expressed as 1.7×10^3 mL to indicate that there are two significant digits. The terminal decimal point is used to clarify the precision of the measurement; 300 mL has one significant digit, but 3.00×10^2 mL, 300. mL and 0.300 L each have three significant digits.

Additional Problems

Determine the number of significant digits (2.1):

0.0040 L   0.125607 mg   53,608 cm^3

0.00000000001215 km   83,800. s   156890000 mL

Significant digits in Calculations

Since significant digits indicate the certainty of the measurement, calculations need to reflect the measurement made with the least amount of certainty. In calculations, you cannot claim more certainty in the form of additional digits. Rounding off is necessary to answer problems with the proper number of significant digits.

Rules for rounding off:

1. If the first non-significant digit is equal to or greater than 5, increase the last significant digit by 1 and drop all non-significant digits. For example, 4.876 rounds to 4.88 if three significant digits are retained and to 4.9 if two significant digits are retained.
2. If the first non-significant digit is less than 5, drop all non-significant digits. For instance, 1.324 rounds to 1.32 if three significant digits are retained and 1.3 if 2 significant digits are retained.
3. If the calculation has multiple steps, carry one of two additional significant digits through the multi-step calculations and round off for the final answer.

Math Operations and Significant Digits:

1. For multiplication or division, the answer contains the same number of significant digits as there are in the measurement with the fewest significant digits. If you wanted to find the area of a rectangle where one side measured 7.38 cm and the other measured 5.2 cm. The area calculation would be

    Area = 7.38 cm x 5.2 cm = 38 cm²

    The calculator may show 38.376 cm², but the answer should be reported with two significant digits as 38 cm², because the measurement with the fewest significant digits, 5.2 cm, contained only two significant digits.

2. For addition or subtraction, the answer must have the same number of decimal places as there are in the measurement with the fewest decimal places. If you wanted to find the mass of three objects that had individual masses of 5g, 15.3g

and 2.38g, you need to look at the number of decimal places of each of the individual masses.

Final Mass = 5 g + 15.2 g + 2.18 g = 22 g

Here the calculator shows 22.38 g, but it is reported as 22 g with zero decimal places as the number in the measurement contained zero decimal places.

Additional Problems (2.2)

What is the density of an unknown metal that has a mass of 13.1g and a volume of 262.5 cm³?

Calculate the following:

$$\frac{63.2g + 78g + 636.25\ g}{7.2\ g}$$

$$\frac{16.7451\ cm - 3.85\ cm}{2.356\ cm}$$

Scientific Notation

Many of the calculation you will perform involve numbers written in exponential notation: $A \times 10^n$, where A, the coefficient, is between is between 1 and 10 ($1 \leq A < 10$) and n is a whole number or integer.

For larger numbers that are greater than 1, n is positive ($n > 0$). For number less than 1, n is negative ($n < 0$). The value of n tells the number of places the decimal point must be moved to be written in conventional notation.

Here's a trick for remembering the sign of the exponents: consider your bank account, if you have more than a dollar that's a "positive" thing, positive exponents. If you have less than a dollar, that's a "negative" thing, negative exponents.

To change a number from conventional to exponential notation, move the decimal point to the left for numbers equal to or greater than 10 and to the right for numbers between 0 and 1.

85,000,000 changes to 8.5×10^7 (decimal point 7 places to the left)

0.001209 changes to 1.209×10^{-3} (decimal point 3 places to the right)

To change a number from exponential to conventional notation, move the decimal point the number of places of the value of n to the right for numbers where n is positive and to the left for numbers where n is negative.

1.59×10^5 changes to 159,000 (decimal point 5 places to the right, n is positive)

8.56×10^{-6} changes to 0.00000856 (decimal point 6 places to the left)

Here are some practice problems to change from exponential to conventional or conventional to exponential (3.1).

63.809   0.0000325068

631,587,598   0.0000000002598

1.235854×10^{-7} 3.2587×10^9

2.3568×10^6 $5.6874298 \times 10^{-14}$

In some types of problems, you will need to perform a mathematical operation on a number that has exponents.

1. To multiply exponential number, multiply the coefficient (the number right in front of the $\times 10^n$), and add the exponents.

   $(5.5 \times 10^2)(8.1 \times 10^5) = (5.5 \times 8.1) \times 10^{(2+5)} = 44.55 \times 10^7 = 4.455 \times 10^8 = 4.5 \times 10^8$

   $(8.3 \times 10^{20})(3.2 \times 10^{-14}) = (8.3 \times 3.2) \times 10^{[20+(-14)]} = 26.56 \times 10^6 = 2.656 \times 10^7 = 2.7 \times 10^7$

2. To divide exponential numbers, divide the coefficients and subtract the exponents.

   $\frac{5.5 \times 10^5}{8.1 \times 10^2} = \frac{5.5}{8.1} \times 10^{(5-2)} = 0.679 \times 10^3 = 6.79 \times 10^2 = 6.8 \times 10^2$

   $\frac{1.5 \times 10^{-5}}{9.2 \times 10^{-7}} = \frac{1.5}{9.2} \times 10^{((-5)-(-7))} = 0.163 \times 10^2 = 1.63 \times 10^1 = 1.6 \times 10^1$

Additional Scientific notation problems (3.2)

$$\frac{(5.2 \times 10^{23}) \times (3.2 \times 10^{-16})}{6.1 \times 10^{-8}} =$$

$$\frac{2.3 \times 10^{-6}}{(9.63 \times 10^{-3})(3.11 \times 10^{-5})} =$$

$$\frac{4 * (3.2 \times 10^{-8})}{9.1 \times 10^{-10}} =$$

Problems Involving pH

pH problems tend to be a little challenging as a *p-scale* is used to account for the enormous range of the acid concentration, $[H_3O^+]$ or $[H^+]$. This is the negative of the common logarithm (base-10) system.

$$pH = -\log [H^+]$$

$$[H^+] = 10^{-pH}$$

Note: $[H_3O^+]$ and $[H^+]$ are interchangeable and mean the same thing: the acid concentration.

Find the pH of a 5.4×10^{-3} M H^+ solution?

$$pH = -\log [H^+] = -\log (5.4 \times 10^{-3}) = 2.267606 = 2.27$$

For pH measurements, the number of decimal places in the pH value is the number of significant figures in the concentration. 5.4×10^{-3} M has two significant figures, so its negative logarithm, 2.27, has two digits to the right of the decimal point or two decimal places.

What is the pH of 10^{-10} *M* H_3O^+ solution?

$$pH = -\log [H_3O^+] = -\log (10^{-10}) = 10$$

Recall that $\log 10^{-x} = -x$, so for the pH equation you need to take the negative log of $[H_3O^+]$, which in this case would equal 10.

Or we can use the second pH formula of $[H^+] = 10^{-pH}$

$$[H^+] = 10^{-10} = 10^{-pH}$$

$$10 = pH$$

Similarly, a 10^{-4} M H_3O^+ solution has a pH of 4 and a 10^{-12} M H_3O^+ solution has a pH of 12.

If the pH of tomato juice is 5, what is its $[H_3O^+]$?

$$[H_3O^+] = 10^{-pH} = 10^{-5} = 0.00001 \text{ M}$$

If the pH of a solution is 3.80, what is its $[H_3O^+]$?

$$[H_3O^+] = 10^{-pH} = 10^{-3.80} = 1.6 \times 10^{-4} \text{ M}$$

The p-scale is also used to express hydroxide ion concentrations.

$$pOH = -\log [OH^-]$$

$$10^{-pOH} = [OH^-]$$

What is the pOH of a NaOH solution with a concentration of 4×10^{-3} M?

$$pOH = -\log [OH^-] = -\log [4 \times 10^{-3}] = 2.4$$

Note: 4×10^{-3} M has one significant digit, so the pOH is expressed with one decimal place

At 25°C, the pOH and pH have a useful relationship.

$$[OH^-][H^+] = 1 \times 10^{-14}$$

$$pOH + pH = 14$$

If we wanted to figure out the pH of the NaOH solution with a concentration of 4×10^{-3} M, we would take the pOH and subtract it from 14.

$$pH = 14 - pOH = 14 - 2.4 = 11.6$$

We can also figure out the H$^+$ concentration by using:

$$[H^+] = \frac{[1 \times 10^{-14}]}{[OH^-]}$$

$$[H^+] = \frac{[1 \times 10^{-14}]}{[4 \times 10^{-3}]} = 0.25 \times 10^{-11} = 2.5 \times 10^{-12}$$

When [H+] < [OH-], pH > 7, the solution is basic

When [H+] > [OH-], pH < 7, the solution is acidic

When [H+] = [OH-], pH = 7, the solution is neutral

Additional pH problems (4.1)

What is the pH of 0.001 M H_2SO_4?

Calculate the pH of 0.00001 M NaOH

Note: pK_a is not the same thing as the pH. The pK_a indicates whether an acid is strong or weak, while the pH indicates whether a system is acidic or basic. As with pH, pK_a are the negative log of the equilibrium constant for the acid dissociation, K_a. pK_b are the negative log of the equilibrium constant for the base dissociation.

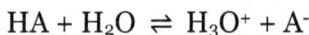

$$HA + H_2O \rightleftharpoons H_3O^+ + A^-$$

$pK_a = - \log K_a$

where $K_a = \frac{[H_3O^+][A^-]}{[HA]}$

$pK_b = -\log K_b$

where $K_b = \frac{[BH^+][OH^-]}{[B]}$

$pK_a + pK_b = 14$

The Henderson-Hasselbalch equation gives the mathematical relationship between pH and pK_a and is used in problems dealing with buffered solutions.

$$pH = pK_a + \log \left(\frac{[base]}{[acid]} \right)$$

This can also be shown as:

$$\frac{[base]}{[acid]} = 10^{pH-pK_a}$$

What is the pH of an Acetic acid buffer system that contains 0.75M $HC_2H_3O_2$ and 0.50M $NaC_2H_3O_2$? The K_a of $HC_2H_3O_2$ is 1.8 x 10⁻⁵.

For this problem, we are given the concentrations of the base and the acid, and the K_a of the Acetic acid in a buffer system. We want to find the pH so we are going to us the following equation:

$$pH = pK_a + log\left(\frac{[base]}{[acid]}\right)$$

We know that the relationship between K_a and pK_a is pK_a = -log K_a so we can find the pK_a by taking the negative log of the K_a.

$$pK_a = -log\,K_a = -log(1.8 \times 10^{-5}) = 4.74$$

Now that we know the pK_a we can plug everything into the equation above to find the pH.

$$pH = pK_a + log\left(\frac{[base]}{[acid]}\right) = 4.74 + log\left(\frac{0.50M}{0.75M}\right)$$

$$pH = 4.57$$

What is the pK_a of an unknown weak acid if the pH of the solution was measured to be 6.32 when [HA] = 0.530M and [A⁻] = 0.872M?

For this problem, we are given the pH and the concentrations of the acid and the base, and we are being asked to find the pK_a. So, in order to do that, we need to rearrange the following equation to solve for pK_a:

$$pH = pK_a + log\left(\frac{[base]}{[acid]}\right)$$

If we subtract log(base/acid) from both sides of the equation, we are left with

$$pK_a = pH - log\left(\frac{[base]}{[acid]}\right)$$

Now we can plug in the values for pH and the concentrations to get the pKₐ.

$$pK_a = 6.32 - log\left(\frac{0.872M}{0.530M}\right)$$

$$pK_a = 6.10$$

Additional problems (4.2):

What is the pH of a solution containing 0.25 mol of NH_4Cl and 2.5 mol of NH_3? The K_b of NH_3 is 1.8×10^{-5}.

Calculate the pH of a solution containing 25g HF and 31g NaF in 750mL of solution. The K_a of HF is 7.2×10^{-4}.

Metric system

The calculations and numbers we use in chemistry represent a specific physical quantity. These quantities include mass, length, time, temperature, and more. There are specific SI Units for these measured and derived quantities. See below.

Table 1. SI Units for Physical Quantities

Physical Quantity	Unit Name	Unit Abbreviation
Mass	kilogram	kg
Length	meter	m
Time	second	s
Amount of substance	mole	mol

Each type of quantity has a base unit. For quantities that are smaller or larger than the base unit, we use prefixes and exponential (scientific) notation. These prefixes are based on powers of 10. The most common prefixes you will come across in chemistry courses are highlighted below in Table 2.

Table 2. Common Decimal Prefixes Used with SI Units

Prefix	Prefix symbol	Number	Exponential Notation
Mega-	M	1,000,000	10^6
Kilo-	K	1,000	10^3
Deca-	da	10	10^1
Deci-	d	0.1	10^{-1}
Centi-	c	0.01	10^{-2}
Milli-	m	0.001	10^{-3}
Micro-	μ	0.000001	10^{-6}
Nano-	n	0.000000001	10^{-9}
Pico-	p	0.000000000001	10^{-12}

Familiarizing yourself with these different quantities will be helpful in dimensional analysis questions, which will be covered in the next section.

Table 2 shows us the value for each prefix. For instance, "kilo-" means 1000; a kilometer is 1000 times longer than a meter. So, 1 kilometer is equal to 1000 meters.

$$1 \; kilometer = 1000 \; meters$$

Or alternatively

$$\frac{1}{1000} \; kilometers = 0.001 \; kilometers = 1 \; meter$$

A microgram (µg) is 1/1000000 the mass of a gram (g). We can write this as

$$1 \; microgram = \frac{1}{1000000} gram = 0.000001 \; grams$$

Or alternatively

$$1000000 \; micrograms = 1 \; gram$$

We can use these equivalences to convert between different units of the same physical quantity. Table 3 breaks down more metric conversions.

Table 3. Common Metric Conversions

Prefix	Meaning	Conversion	
pico (p)	1/1000000000000	1000000000000 pico_____	= 1 _____
nano (n)	1/1000000000	1000000000 nano_____	= 1 _____
micro (µ)	1/1000000	1000000 micro_____	= 1 _____
milli (m)	1/1000	1000 milli_____	= 1 _____
centi (c)	1/100	100 centi_____	= 1 _____
deci (d)	1/10	10 deci_____	= 1 _____
deca (da)	10	1 deca_____	= 10 _____
kilo (k)	1000	1 kilo _____	= 1000_____
mega (M)	1000000	1 mega _____	= 1000000 _____

How many centimeters are there in 1 meter?

The first thing to do is to recall what the prefix centi means. In looking at Table 3, we see that centi is 1/100. From this we know that 100 centimeters is equal to 1 meter

$$100 \; centimeters = 1 \; meter$$

Or alternatively if we divide both sides of the equation by 100,

$$1 \; centimeter = \frac{1}{100} meter = 0.01 \; meter$$

Now, if you are asked to find the number of centimeters in 1 kilometer, we would need to do a two-part conversion.

First, we need to convert centimeters (cm) to meters (m), then convert meters (m) to kilometers (km).

Game plan: cm → m → km

As we discussed, "centi-" means 1/100 so we know that 100 centimeters is equal to 1 meter.

$$100 \; cm = 1m$$

Now we need to convert meters (m) to kilometers. We know that kilo means 1000 so we know that 1 kilometer is 1000 meters.

$$1 \; km = 1000 \; m$$

Or alternatively

$$1 \; m = \frac{1}{1000} km = 0.0001 \; km$$

So if we combine these equations together, we find

$$100 \; cm = 1 \; m = 0.0001 \; km$$

Which can be rewritten as

$$100 \; cm = 0.0001 \; km$$

If we multiply both sides of the equation by 1000

$$100000 \; cm = 1 \; km$$

We find that there are 100,000 cm in 1 km.

Alternatively, we can find that

$$1cm = 0.00001 \; km$$

Additional problems (5.1)

How many seconds are in 1 nanosecond? In decisecond?

How many picograms are in 1 gram? In 1 kilogram?

How many milliliters are in 1 liter? In 1 megaliter?

Dimensional Analysis

All measured quantities consist of a number and a unit; a candy bar is 140 calories, and a person is 6 feet tall. Ratios of quantities include ratios of units. For instance, your car speed is measured in miles/hour. When doing chemistry problems, it is important that you write your units as they will help you figure out what mathematical operation you need to do.

For example, if you wanted to know the speed in units of miles/hour of a car traveling 220 miles in 4 hours, you can use the units to set up the equation to solve the problem.

$$Speed \left(\frac{miles}{hour}\right) = \frac{220\ miles}{4\ hours} = 55\ \frac{miles}{hour}$$

To figure this problem out you would take the miles and divide by the hours so that your final unit is in terms of miles/hour. When you take 220 miles and divide it by 4 hours, you get a speed of 55 miles/hour. The answer 55 miles/hour can be understood as 55 miles over 1 hour.

However, you are not necessarily given the correct units to solve a problem. What if, for example, you were asked to determine the speed in miles per hour of a car traveling 220 miles in 240 minutes. You would need to change the units of time from minutes to hours. This type of conversion is referred to as dimensional analysis.

In dimensional analysis, the given unit is multiplied by one (or more) conversion factors to get the desired unit in the answer. **Conversion factors** are ratios used to express measured quantities in different units. You will need to make sure your conversion factors have the correct units to get the desired answer.

Conversion factors are equivalent to one. They are necessary for unit conversions. We can construct conversion factors through equivalency. For instance, 1 foot is equal to 12 inches.

$$1\ foot = 12\ inches$$

We can construct two conversion factors from this equivalency by division. If we divide both sides of the equation by 1 foot, the conversion factor we obtain is:

$$\frac{1 \ foot}{1 \ foot} = \frac{12 \ inches}{1 \ foot} = 1$$

If we divide both sides of the equation by 12 inches, the conversion factor we obtain is:

$$\frac{1 \ foot}{12 \ inches} = \frac{12 \ inches}{12 \ inches} = 1$$

When doing the dimensional analysis problems, using the conversion factor is the same as multiplying by 1. The number and unit of the quantity change, but the size of the quantity does not.

When choosing a conversion factor, you want to select a factor where your given unit cancels and you are left with your desired unit.

$$given \ unit \ \times \left(\frac{desired \ unit}{given \ unit}\right) = desired \ unit$$

↑

Conversion Factor

If we wanted to figure out how many inches are in 3.7 ft. We would use the conversion factor with inches in the numerator and feet in the denominator.

$$3.7 \ ft \ \times \frac{12 \ in}{1 \ ft} = 44 \ in$$

If we wanted to figure out the speed in mph of a car traveling 220 miles in 240 minutes, we would need to convert minutes to hours. We know that there are 60 minutes in 1 hour so we can use this equivalency to determine a conversion factor.

$$60 \ minutes = 1 \ hour$$

If we divide both sides of the equation by 60 minutes we get

$$\frac{60 \ minutes}{60 \ minutes} = \frac{1 \ hour}{60 \ minutes} = 1$$

So for this problem, the conversion factor we will use is $\frac{1 \ hour}{60 \ minutes}$:

$$240 \ minutes \ \times \ \frac{1 \ hour}{60 \ minutes} = 4 \ hours$$

$$\uparrow \qquad\qquad \uparrow \qquad\qquad \uparrow$$

given units   conversion factor desired units

Now we can use the time of 4 hours to determine the mph of the car.

$$Speed \left(\frac{miles}{hour}\right) = \frac{220 \ miles}{4 \ hours} = 55 \ \frac{miles}{hour}$$

If you wanted to figure out how many moles are in 3.0 g of carbon, you would need to convert g to moles.

It is helpful to make a quick game plan of what units you have and which units you want to get by using arrows like: g → moles

$$3.0 \ g \ \times \ \frac{\_\_\_ \ mol \ Carbon}{\_\_g \ Carbon} = \_\_\_ \ mol \ Carbon$$

The conversion factor in this example would need to contain moles in the numerator and grams in the denominator such that the unit of grams would cancel out and you would be left with moles for your answer.

$$3.0 \ \cancel{g} \ \times \ \frac{\_\_\_ \ mol \ Carbon}{\_\_\cancel{g} \ Carbon} = \_\_\_ \ mol \ Carbon$$

Using the periodic table, we can find the conversion factor using the atomic weight of carbon. The atomic weight of Carbon is 12 g and we know that the atomic weight is based on 1 mol.

$$12 \ g \ Carbon = 1 \ mol \ Carbon$$

If we divide both sides of the equation by 12 g Carbon we obtain:

$$\frac{12 \ g \ Carbon}{12 \ g \ Carbon} = \frac{1 \ mol \ Carbon}{12 \ g \ Carbon} = 1$$

We can now use $\frac{1 \, mol \, Carbon}{12 \, g \, Carbon}$ as our conversion factor.

$$3.0 \, \cancel{g} \times \frac{1 \, mol \, Carbon}{12 \, \cancel{g} \, Carbon} = 0.25 \, mol \, Carbon$$

↑   ↑   ↑

given units   *conversion factor*   *desired units*

We found that there are 0.25 moles of Carbon in 3.0 grams of Carbon.

Similarly, we can convert from moles to grams of an element. Say for example, you want to find the mass in grams of Iron in 0.234 moles. To solve for this, we need to have a conversion factor with grams in the numerator and moles in the denominator so we are left with our answer being in moles.

$$0.234 \, mol \, Fe \times \frac{\_ \, g \, Fe}{\_ \, mol \, Fe} = g \, Fe$$

Based on the periodic table, we see that the atomic mass of Fe is 55.8 per mole of Fe.

$$55.8 \, g \, Fe = 1 \, mol \, Fe$$

If we divide both sides by mol Fe we get:

$$\frac{55.8 \, g \, Fe}{1 \, mol \, Fe} = \frac{1 \, mol \, Fe}{1 \, mol \, Fe} = 1$$

Using $\frac{55.8 \, g \, Fe}{1 \, mol \, Fe}$ as a conversion factor we can now determine how many grams are in 0.234 mol Fe.

$$0.234 \, mol \, Fe \times \frac{55.8 \, g \, Fe}{1 \, mol \, Fe} = 13.1 \, g \, Fe$$

There are 13.1 g Fe in 0.234 mol Fe.

Additional problems 6.1

What is the price of a piece of 325 cm long copper wire that costs $0.90/ft? [1 in = 2.54 cm]

The radius of a copper atom is 128pm. What is its radius in nanometers (nm)?

How many grams are in 0.428 moles of sodium bicarbonate, NaHCO3?

Stoichiometry

Stoichiometry comes from the Greek words "stoikheîon," meaning "element," and "métron," meaning "measure." The majority of stoichiometry problems you will do in chemistry involve some degree of dimensional analysis. When doing these problems, writing down units is very important, and will often times help you figure out what mathematical operation you need to carry out to solve the problem.

Some stoichiometric problems may require multiple conversion factors. For instance, if we are asked how many hydrogen atoms are there in 2.50g of NH_3? We need to convert g NH_3 to moles NH_3 then to molecules of NH_3 then finally to atoms of H.

A helpful strategy in solving stoichiometry problems is to make a flow chart, or game plan, that shows what conversions need to be carried out to solve the problem. For this example, our game plan would be:

$$g\ NH_3 \rightarrow moles\ NH_3 \rightarrow molecules\ NH_3 \rightarrow atoms\ of\ H$$

Step 1: To go from g → moles, the conversion factor will have g in the denominator and moles in the numerator

Conversion factor 1: $\frac{moles}{g}$

Looking at the periodic table, we find that the mass of NH_3 is 17.0g/mol (N: 14 g/mol and 3 x H: 1 g/mol) so to write this in terms of $\frac{moles}{g}$, it would be $\frac{1\ mol}{17.0\ g}$

Step 2: To go from moles → molecules, the conversion factor will have moles in the denominator and molecules in the numerator

Conversion factor 2: $\frac{molecules}{moles}$

The conversion from molecules to moles includes Avogadro's number, which is 6.02 x 10²³ molecules in 1 mole. So our second conversion factor would be $\frac{6.02 \times 10^{23}\ molecule}{1\ mol}$

Step 3: To go from molecules → atoms H, the conversion factor will have molecules in the denominator and atoms in the numerator

Conversion factor 3: $\frac{atoms}{molecules}$

Each molecule of NH_3, contains 3 atoms of Hydrogen and 1 atom of Nitrogen, so our third conversion factor would be $\frac{3\ H\ atoms}{1\ NH_3\ molecule}$

With these conversion factor units determined, let's start the problem.

$$2.50\ g\ NH_3\ \times\ \frac{1\ mol}{17.0g}\ \times\ \frac{6.02\times10^{23}molecule}{1\ mol}\ \times\ \frac{3\ H atoms}{1\ NH_3 molecule}$$

Conversion factor   1   2   3

$$=\ 2.66\ x\ 10^{23}\ H\ atoms$$

Other dimensional analysis problems, may involve chemical equations such as:

How many moles of H_2O are produced from 2.15 moles of O_2, given the following equations?

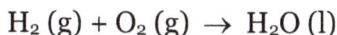

$$H_2\ (g) + O_2\ (g) \rightarrow H_2O\ (l)$$

The first thing we need to do before we begin any type of calculations, is determine if the equation is balanced. To do this, you need to count the number of elements on each side. In this example, the reactants contain 2 Hydrogens and 2 Oxygens and the product contains 2 Hydrogens and 1 Oxygen. Since the reactants contain more oxygen atoms than the product, we see that the equation is not balanced. To balance the equation, we will need to use coefficients.

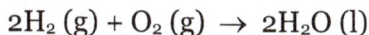

$$2H_2\ (g) + O_2\ (g) \rightarrow 2H_2O\ (l)$$

By placing a 2 in front of the H_2 and a 2 in front of the H_2O, we now have a balanced equation. The coefficients indicate that 2 molecules of H_2 react with each molecule of O_2 to form 2 molecules of H_2O. These stoichiometric relationships can be used to convert moles of O_2 to moles of H_2O as asked in our original problem. For tips on balancing equations or further instructions, consult your chemistry text book.

We are given 2.15 moles of O_2 and are being asked how many moles of H_2O are produced. So, we will need to covert moles O_2 to moles H_2O.

moles O_2 → moles H_2O

The conversion factor for this problem will contain moles of O_2 in the denominator and moles of H_2O in the numerator. The coefficients of the equation will be the numerical values for the conversion factor, as 1 molecule of O_2 produces 2 molecules of H_2O.

$$2.15 \; mol \; O_2 \; x \; \frac{2 \; mol \; H_2O}{1 \; mol \; O_2} = 4.30 \; mol \; H_2O$$

Consider the combustion of butane:

$$4 \; C_4H_{10} + 13 \; O_2 \rightarrow 8 \; CO_2 + 10 \; H_2O$$

How much CO_2 is produced when 1.0 g of C_4H_{10} is burned?

First thing: is the chemical reaction balanced? Yes, now we can move on to conversions. To solve this problem, we will first need to convert 1.0 g of C_4H_{10} to moles, then convert moles C_4H_{10} to moles of CO_2 using their coefficients, and then finally convert moles CO_2 to grams CO_2 using molar mass of CO_2.

Game plan: g C_4H_{10} → mol C_4H_{10} → mol CO_2 → g CO_2

$$1.0 \; g \; C_4H_{10} \; \times \frac{1 \; mol \; C_4H_{10}}{58.0 \; g \; C_4H_{10}} \times \frac{8 \; mol \; CO_2}{4 \; mol \; C_4H_{10}} \times \frac{44.0 \; g \; CO_2}{1 \; mol \; CO_2} = 1.5 \; g \; CO_2$$

molar mass of C_4H_{10} coefficients in eq   molar mass of CO_2

We can also calculate the amount of O_2 consumed and H_2O produced from this reaction.

For the amount of O_2 consumed, we would need to covert convert 1.0g of C_4H_{10} to moles, then convert moles C_4H_{10} to moles of O_2 using their coefficients, and then finally convert moles O_2 to grams O_2 using molar mass of O_2.

Game plan: g C_4H_{10} → mol C_4H_{10} → mol O_2 → g O_2

$$1.0 \; g \; C_4H_{10} \; \times \frac{1 \; mol \; C_4H_{10}}{58.0 \; g \; C_4H_{10}} \times \frac{13 \; mol \; O_2}{4 \; mol \; C_4H_{10}} \times \frac{32.0 \; g \; O_2}{1 \; mol \; O_2} = 1.8 \; g \; O_2$$

Additional Stoichiometry problems (7.1)

Figure out how much H_2O is produced from the combustion of 1.0g of butane.

Sodium bicarbonate and citric acid react together in an acid base reaction as follows:

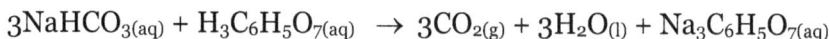

$$3NaHCO_{3(aq)} + H_3C_6H_5O_{7(aq)} \rightarrow 3CO_{2(g)} + 3H_2O_{(l)} + Na_3C_6H_5O_{7(aq)}$$

If 1.21 grams of sodium bicarbonate and excess citric acid react, how many grams of CO_2 form?

A helpful general approach to these types of problems is as follows:

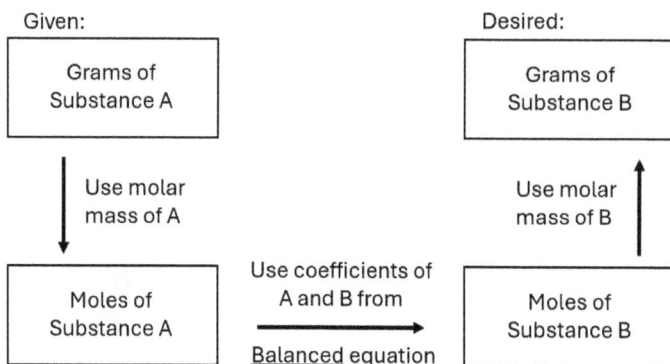

Given:

| Grams of Substance A |

Use molar mass of A

| Moles of Substance A |

Use coefficients of A and B from
→
Balanced equation

Desired:

| Grams of Substance B |

Use molar mass of B

| Moles of Substance B |

Figure 2. Basic Stoichiometry Flow Chart

Limiting Reagent

When the quantity of one reactant is given, it is initially assumed that the amount of the other reactants is sufficient to fully produce the product. However, one reactant may not be as present as much as the other reactants. This reactant limits the amount of product

35

that can be produced and is referred to as the limiting reagent or limiting reactant.

We come across limiting reactants in everyday life. Consider hotdogs: the number of hotdogs that can be made from one pack of hotdogs and one pack of buns depends on the relative quantities in each pack. Hot dogs come in packs of 10, and hotdog buns come in packs of 8. So, if you had one pack of each, you could only make 8 hotdogs since there are only 8 buns. The buns would be the limiting reagent.

Consider the following,

$$2 NH_3 + 3 CuO \rightarrow N_2 + 3Cu + 3H_2O$$

If 18.1 g of NH_3 are reacted with 90.4g of CuO, which is the limiting reagent? How many grams of N_2 will form?

The first thing that needs to be done is convert g to moles of each of the reactant. Since we are being asked to figure out how much N_2 will form, the next step will be to convert moles of reactant (NH_3 and CuO) to moles of N_2 using the coefficients of the balanced equation. From there we can convert moles of N_2 to g of N_2 using the molar mass of N_2.

Game plan:

$$g\ CuO \rightarrow mol\ CuO \rightarrow mol\ N_2 \rightarrow g\ N_2$$

AND

$$g\ NH_3 \rightarrow mol\ NH_3 \rightarrow mol\ N_2 \rightarrow g\ N_2$$

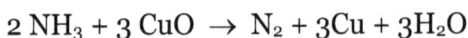

$$18.1\ g\ NH_3 \times \frac{1\ mol\ NH_3}{17.0g\ NH_3} \times \frac{1\ mol\ N_2}{2\ mol\ NH_3} \times \frac{28.0\ g\ N_2}{1\ mol\ N_2} = 14.9\ g\ N_2$$

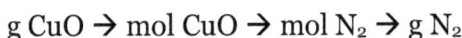
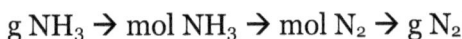

$$90.4\ g\ CuO \times \frac{1\ mol\ CuO}{17.0g\ CuO} \times \frac{1\ mol\ N_2}{3\ mol\ CuO} \times \frac{28.0\ g\ N_2}{1\ mol\ N_2} = 10.6g\ N_2$$

Based on the above calculations, the 18.1 g NH_3 produces 14.9 g N_2 and the 90.4 g CuO produces 10.6 g N_2. Since the 90.4 g CuO produces less N_2, it is the limiting reactant, and the amount of N_2 that can be produced from these quantities of reactants is 10.6g.

Consider the following balanced equation:

$$2N_2H_4 + N_2O_4 \rightarrow 3N_2 + 4H_2O$$

How many grams of N_2 form when 131 g of N_2H_4 and 245 g of N_2O_4 are combined?

To solve this problem, we need convert g of the reactants to moles using the molar masses. Then convert those moles to moles of N_2 using the coefficients of the balanced equation. The moles of N_2 can then be converted to grams of N_2 by multiplying with the molar mass of the nitrogen.

Game plan:

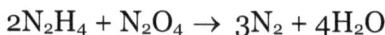

$$\text{g } N_2H_4 \rightarrow \text{mol } N_2H_4 \rightarrow \text{mol } N_2 \rightarrow \text{g } N_2$$

AND

$$\text{g } N_2O_4 \rightarrow \text{mol } N_2O_4 \rightarrow \text{mol } N_2 \rightarrow \text{g } N_2$$

$$131 \text{ g } N_2H_4 \times \frac{1 \text{ mol } N_2H_4}{32.1 \text{g } N_2H_4} \times \frac{3 \text{ mol } N_2}{2 \text{ mol } N_2H_4} \times \frac{28.0 \text{ g } N_2}{1 \text{ mol } N_2} = 171 \text{ g } N_2$$

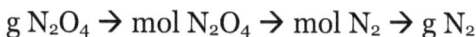

$$245 \text{ g } N_2O_4 \times \frac{1 \text{ mol } N_2O_4}{92.0 \text{g } N_2O_4} \times \frac{3 \text{ mol } N_2}{1 \text{ mol } N_2O_4} \times \frac{28.0 \text{ g } N_2}{1 \text{ mol } N_2} = 224 \text{ g } N_2$$

Based on the above calculations, 131 g N_2H_4 generates 171 g N_2, whereas 245 g N_2O_4 produces 224 g N_2. Since 171 g < 224 g, we can deduce that N_2H_4 is the limiting reagent, and that only 171 g of N_2 will be produced from this reaction.

The common mistakes that are made with limiting reagent problems, is choosing the reagent with less moles rather than comparing the moles of the product produced. The second common problem is to choose the reactant with the lower mass as the limiting reactant, rather than computing the mass of the product formed from both quantities of the reactants. By taking the quantities of each reactant and doing the dimensional analysis to figure out the product quantity these two mistakes can be avoided.

Additional limiting reagent problems (7.2)

How many grams of solid aluminum sulfide can be prepared by the reaction of 5.0 g of aluminum and 10.0 g sulfur?

Sodium hydroxide reacts with carbon dioxide as follows:

$2NaOH_{(s)} + CO_{2(g)} \rightarrow Na_2CO_{3(s)} + H_2O_{(l)}$

If 1.75 mol NaOH and 2.50 mol CO_2 are allowed to react, what is the limiting reactant? How many moles of water will be produced?

How much molten iron is produced from the reaction of 35.0 g FeO with 25.0 g Al? Given the balanced equation: $3FeO_{(l)} + 2Al_{(l)} \rightarrow 3Fe_{(l)} + Al_2O_{3(s)}$

Percent Calculations

Percent calculations appear in a few areas of chemistry, including percent composition, percent ionization, percent concentration, and percent yield. A percent (%) expresses the amount of a part compared to a whole.

$$\frac{part}{whole} \times 100\% = \%$$

If we wanted to find the percent a quarter is to a dollar, we would need to divide the part, the quarter, by the whole, a dollar and multiply that ratio by 100%.

$$\frac{25\ cents}{100\ cents} \times 100\% = 25\%$$

We see that a quarter is 25% of a dollar.

Percent Composition

For percent composition problems, we follow this same equation. For example, sterling silver contains both silver and copper metals. If a sterling silver chain contains 21.5 g of silver and 2.3 g of copper, what is the percent of silver?

To solve this, we need to take the part that is silver and divide that by the entire sterling silver material which is the sum of both metals then multiply that ratio by 100%.

$$\frac{21.5\ g}{(21.5 + 2.3)\ g} \times 100\% =$$

$$\frac{21.5\ g}{23.8\ g} \times 100\% = 90.3\ \%$$

We find that the sterling silver chain contains 90.3% silver.

An 18-karat gold ring contains 10.7g of gold, 1.3g of silver and 2.1g of copper. What is the percent of gold in the 18-karat ring?

$$\frac{g\ gold}{total\ g\ of\ ring} \times 100\% =$$

$$\frac{10.7g}{(10.7 + 1.3 + 2.1)\ g} \times 100\% =$$

$$\frac{10.7\ g}{14.1\ g} \times 100\% = 75.9\%$$

Other types of percent composition questions could look something like: calculate the percent carbon by mass in $C_6H_{12}O_6$, where you are not given any quantities in grams. To solve these types of problems, you will need to use the atomic and formula weights. The equation we use is still:

$$\frac{part}{whole} \times 100\% = \%$$

Now the part portion will contain the number of atoms of that element multiplied by the atomic weight of the element to give the entire part that element occupies. And the whole portion will be the formula weight of that molecule or substance.

For example, calculate the percent carbon by mass in $C_6H_{12}O_6$, we will need to take the mass of carbon and multiply it by 6 as that is how many carbons are in the entire sample then divide by the formula mass $C_6H_{12}O_6$.

$$\frac{(mass\ of\ Carbon)(\#\ of\ Carbons)}{mass\ of\ C_6H_{12}O_6} \times 100\% =$$

To determine the mass of $C_6H_{12}O_6$, we need to take the mass of each element and multiply it by the number of elements contained in the molecule.

$$\frac{(12\ amu)(6)}{(12\ amu)(6) + (1\ amu)(12) + (16\ amu)(6)} \times 100\% =$$

There are 6 carbons which have a mass of 12 amu each, so the portion of the molecule from carbon is 6 × 12 amu. The 12 hydrogens have a mass of 1 amu each, so the contribution to the mass from hydrogen is 12 × 1 amu. The 6 oxygens have a mass of 16 amu each, oxygen contributes 6 × 16 amu to the formula weight.

$$\frac{72 amu}{72\ amu + 12\ amu + 96\ amu} \times 100\% =$$

The carbon portion is 72 amu, the hydrogen portion is 12 amu and the oxygen portion is 96 amu.

$$\frac{72\ amu}{180\ amu} \times 100\% = 40\%\ Carbon$$

The overall formula weight for $C_6H_{12}O_6$ is 180 amu, and the percent of carbon is 40%.

For additional practice (8.1)

What is the percent hydrogen and the percent oxygen of $C_6H_{12}O_6$?

What is the percent nitrogen, by mass, in $Ca(NO_3)_2$?

What is the percent aluminum, by mass, in $Al(OH)_3$?

Percent Ionization

The percent ionization is another measure of acid strength. The stronger the acid, the greater the percent ionization. Like with the other percent calculations, percent ionization includes the ionized concentration part divided by the concentration of the whole multiplied by 100.

$$Percent\ ionization = \frac{concentration\ ionized}{originial\ concentration} \times 100\%$$

For any acid, the concentration of acid that ionizes is equal to the concentration of $H^+_{(aq)}$ that forms.

$$Percent\ ionization = \frac{[H^+]\ equilibrium}{[HA]\ initial} \times 100\%$$

For example, a 0.047 M solution of HCOOH contains 3.5 x 10^{-3} M $H^+_{(aq)}$ and its percent ionization is:

$$Percent\ ionization = \frac{[H^+]\ equilibrium}{[HA]\ initial} \times 100\%$$

$$= \frac{3.5\ x\ 10^{-3}\ M}{0.047\ M} \times 100\% = 7.4\%$$

Another example is for a 0.10 M solution of nitrous acid (HNO_2) that contains 2.4 x 10^{-3} M $H^+_{(aq)}$. Calculate the percentage of acid that has ionized.

$$Percent\ ionization = \frac{[H^+]\ equilibrium}{[HA]\ initial} \times 100\%$$

$$= \frac{2.4\ x\ 10^{-3}\ M}{0.10\ M} \times 100\% = 2.4\%$$

Consider a 0.030 M solution of ascorbic acid that has a pH of 3.26. Calculate the percent ionization of the ascorbic acid.

Recall that $[H^+] = 10^{-pH}$, so $[H^+] = 10^{-3.26} = 5.46$ x 10^{-4} M

$$Percent\ ionization = \frac{[H^+]\ equilibrium}{[HA]\ initial} \times 100\%$$

$$= \frac{5.46\ x\ 10^{-4}\ M}{0.030\ M} \times 100\% = 1.83\%$$

Additional practice problems (8.2)

A 0.020 M solution of niacin has a pH of 2.76. Calculate the percent ionization of the niacin.

A 0.010 M solution of H_3PO_4 has contains 3.7 x 10^{-3} M $H^+_{(aq)}$. Calculate the percentage of acid that has ionized.

Percent Concentration

In addition to Molarity (moles of a solute/liters of a solution), concentrations of solutions are given as mass percent, % (m/m), volume percent, % (v/v), and mass by volume percent, % (m/v). Like with the other percent calculations, we take the ratio of the part over the whole and multiply it by 100%.

For mass percent, % (m/m), we take the mass of the solute and divide it by the mass of the solution (solute plus the solvent) and multiply it by 100.

$$Mass\ percent = \frac{mass\ of\ solute}{mass\ of\ solute + mass\ of\ solvent} \times 100\%$$

$$= \frac{mass\ of\ solute}{mass\ of\ solution} \times 100\%$$

It is important to remember that the mass of the solute and solvent are additive, meaning that the mass of the solute needs to be added to the mass of the solvent for this type of calculation. Sometimes you will see this written as % (w/w) for ration of weights.

For volume percent, % (v/v), we take the volume of the solute and divide it by the volume of the solution and multiply it by 100.

$$Volume\ percent = \frac{volume\ of\ solute}{volume\ of\ solution} \times 100\%$$

For the volume percent, the volume of the solution is the total volume. Adding the volume of the solute is not needed as it is already incorporated to the total volume of the solution.

For mass by volume percent, % (m/v), we take the mass of the solute and divide it by the volume of the solution and multiply by 100.

$$mass\ by\ volume\ percent = \frac{mass\ of\ solute}{volume\ of\ solution} \times 100\%$$

Consider a 12 oz can of seasonal ale contains 11% alcohol by volume. How many ounces of alcohol does the ale contain?

Here we are given the volume % and the volume of the solution. We are being asked to find the volume of the solute alcohol.

We need to rearrange the volume percent equation to solve for the volume of the solute.

$$Volume\ percent = \frac{volume\ of\ solute}{volume\ of\ solution} \times 100\%$$

Is rearranged to:

$$volume\ of\ solute = \frac{volume\ percent\ \times\ volume\ solution}{100\%}$$

Now we can plug in our values for the volume % and volume of solution to get:

$$volume\ of\ solute = \frac{11\ \times\ 12oz}{100\%} = 1.32\ oz$$

Here's another example. A simple syrup was made by combining 200 g of sucrose with 240 g of water. Calculate the mass percent of sucrose.

For this problem we are given the mass of sucrose and the mass of water for the mixture. We need to calculate the percent mass of sucrose. Using the mass percent equation we can simply plug in the values to obtain the % mass of sucrose.

$$Mass\ percent = \frac{mass\ of\ solute}{mass\ of\ solute + mass\ of\ solvent} \times 100\%$$

$$Mass\ percent = \frac{200\ g\ sucrose}{200\ g\ sucrose + 240\ g\ water} \times 100\%$$

$$Mass\ percent = \frac{200\ g}{440\ g} \times 100\% = 45\%$$

One more example: A glucose solution, $C_6H_{12}O_6$, is prepared by adding 10.0 g of glucose to enough water to make a 500.0 mL solution. What is the % (m/v) of this glucose solution?

Here we are given the mass of the solute, glucose, and the volume of the solution. We are being asked to find the % (m/v) of the solution. To do that we will simply plug in the values we are given into the % (m/v) equation.

$$mass\ by\ volume\ percent = \frac{mass\ of\ solute}{volume\ of\ solution} \times 100\%$$

$$mass\ by\ volume\ percent = \frac{10.0\ g}{500.0\ mL} \times 100\%$$

$$mass\ by\ volume\ percent = 2.00\% \left(\frac{m}{v}\right)$$

Additional problems (8.3)

$CaCl_2$ is used to melt ice on roads. A 25.6 g sample was evaporated leaving a 4.36 g residue. What is the % (m/m) of the sample assuming the residue contained only $CaCl_2$?

What volume of a glucose solution would contain 0.185 g given the solution is 5.0% (m/v)?

Commercially available rubbing alcohol contains 70% isopropyl alcohol by volume. How many mL of isopropyl are in a 500.0 mL container?

Percent Yield

Another percent calculation you will encounter in chemistry is % yield calculations. This is the ratio of the amount of product that was produced to that of the theoretical yield (the amount of product that could form when all of the limiting reactant is consumed) multiplied by 100%. The equation for % yield is:

$$Percent\ Yield = \frac{actual\ yield}{theorectical\ yield} \times 100\%$$

The process of converting iron ore into iron is as follows:

$$Fe_2O_{3(s)} + 3CO_{(g)} \rightarrow 2Fe_{(s)} + 3CO_{2(g)}$$

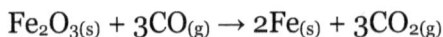

Assuming all 175 g of Fe_2O_3 was consumed as the limiting reagent, what is the theoretical yield of Fe? If 102 g of Fe was obtained experimentally, what is the percent yield?

The first step in this problem is to convert grams of Fe_2O_3 to moles of Fe_2O_3, then convert to moles of Fe, and finally convert to grams of Fe. (Figure 2)

Game plan: g Fe_2O_3 → mol Fe_2O_3 → mol Fe → g Fe

$$175 \; g \; Fe_2O_3 \; \times \frac{1 \; mol \; Fe_2O_3}{159.7 \; g \; Fe_2O_3} \times \frac{2 \; mol \; Fe}{1 \; mol \; Fe_2O_3} \times \frac{55.9 \; g \; Fe}{1 \; mol \; Fe} = 123 \; g \; Fe$$

We are given 175 g Fe_2O_3, which we convert to moles using the FW of 159.7g/mol Fe_2O_3. We then use the equation coefficients to convert to mol of Fe and then use the atomic weight of Fe of 55.9g/mol to convert to grams of Fe. The 122.5 g Fe is the theoretical yield, the maximum amount of Fe that will form when all the 175 g of Fe_2O_3 is consumed.

To determine the percent yield we take the experimental or actual yield and divide it by the theoretical yield and multiply by 100%.

$$Percent \; Yield = \frac{actual \; yield}{theorectical \; yield} \times 100\%$$

$$Percent \; Yield = \frac{102 \; g}{123 \; g} \times 100\% = 82.9\%$$

Here's a problem. Calcium nitrate and ammonium fluoride react to form calcium fluoride, dinitrogen monoxide, and water vapor. What is the mass of calcium fluoride after 15.3 g of calcium nitrate react completely? What is the % yield of calcium fluoride, if 6.30 g are formed experimentally?

You may get a problem like the one above to test your knowledge on chemical nomenclature. This type of question where you need to know the formulas of the chemicals based on the names is a common test giving strategy. It tests the students' ability to write the formulas correctly and do the appropriate math steps to answer the question.

Reactants: Calcium nitrate: $Ca(NO_3)_2$, Ammonium fluoride: NH_4F

Products: Calcium fluoride: CaF_2, Dinitrogen monoxide: N_2O, Water: H_2O

Reaction: $Ca(NO_3)_2 + NH_4F \rightarrow CaF_2 + N_2O + H_2O$

Before we begin with the g → mol conversions, we must make sure that the equation is balanced. In looking at the above reaction, we see that we do not have a balanced equation. To balance the equation we need to put coefficients in front of ammonium fluoride, dinitrogen monoxide and water.

$$Ca(NO_3)_2 + 2NH_4F \rightarrow CaF_2 + 2N_2O + 4H_2O$$

Now with the balanced equation we can begin our calculations. We are given g $Ca(NO_3)_2$ and being asked to find the mass of CaF_2 that is formed.

Game plan: g $Ca(NO_3)_2$ → mol g $Ca(NO_3)_2$ → mol CaF_2 → g CaF_2

$$15.3\ g\ Ca(NO_3)_2 \times \frac{1\ mol\ Ca(NO_3)_2}{164.1\ g\ Ca(NO_3)_2} \times \frac{1\ mol\ CaF_2}{1\ mol\ Ca(NO_3)_2} \times \frac{78.1\ g\ CaF_2}{1\ mol\ CaF_2}$$
$$= 7.28 g\ CaF_2$$

We find that 7.28 g CaF_2 are formed when all 15.3 g $Ca(NO_3)_2$ are consumed. We will use this value as the theoretical yield to find the % yield based. We are given the experimental yield of 6.30 g. So the % yield is as follows:

$$\% \ yield = \frac{6.30\ g}{7.28\ g} \times 100\% = 86.5\%$$

Here's another example. When benzene reacts with bromine, bromobenzene is produced.

$$C_6H_6 + Br_2 \rightarrow C_6H_5Br + HBr$$

If the reaction has 75.0% yield of bromobenzene, how many grams of benzene are needed to make 20.0 g C_6H_5Br?

This question is much trickier than the previous questions as the actual and percent yields are given, you need to solve for the theoretical yield of bromobenzene then convert the grams of bromobenzene to grams benzene through molar conversions.

$$Percent\ Yield = \frac{actual\ yield}{theorectical\ yield} \times 100\%$$

We know that the percent yield equation can be rearranged by cancelling. So to solve for the theoretical yield we will first need to divide both sides by 100%.

$$\frac{percent\ yield}{100\%} = \frac{actual\ yield}{theorectical\ yield}$$

To isolate the theoretical yield, we need to multiply each side of the equation by the theoretical yield.

$$\left(\frac{percent\ yield}{100\%}\right) \times theorectical\ yield = actual\ yield$$

Finally, we can then divide each side of the equation by percent yield/100 and we are left with:

$$theorectical\ yield = \frac{actual\ yield}{\left(\frac{percent\ yield}{100\%}\right)}$$

Which can be re-written as:

$$theorectical\ yield = \frac{actual\ yield \times 100\%}{percent\ yield}$$

For this particular example, we know that the % yield is 75.0%, and that the actual yield is 20.0g for C_6H_5Br. If we plug these values in to our rearranged formula we find that

$$theorectical\ yield = \frac{20.0g \times 100}{75.0} = 26.7\ g$$

Now we know the g C_6H_5Br and we can convert that to mol C_6H_5Br using the MM of C_6H_5Br then convert to mol C_6H_6 using the

48

coefficients of the balanced equation, then to grams of C_6H_6 using the MM of C_6H_6.

Game plan: g C_6H_5Br → mol C_6H_5Br → mol C_6H_6 → g of C_6H_6

$$26.7\ g\ C_6H_5Br\ \times\ \frac{1\ mol\ C_6H_5Br}{157.0\ g\ C_6H_5Br}\times\frac{1\ mol\ C_6H_6}{1\ mol\ C_6H_5Br}\times\frac{78.1\ g\ C_6H_6}{1\ mol\ C_6H_6}$$

$$=13.3\ g\ C_6H_6$$

Additional % yield problems (8.4)

What is the percent yield of a reaction where 230 g of phosphorus trichloride reacts with excess water to form 152 g HCl and aqueous phosphorous acid (H_3PO_3)?

The combustion of butane, C_4H_{10}, a component of natural gas, proceeds as follows:

$$2C_4H_{10} + 13O_2 \rightarrow 8CO_2 + 10H_2O + heat$$

If 10 g of butane produce 13.3 g of CO_2, what is the percent yield of CO_2?

When 56.6 g sulfur and 30.5 g oxygen gas undergo a reaction that has a 90.0% yield, what mass of sulfur dioxide forms?

Crystal Structure Geometry

For purposes of this text, we will be focusing on the cubic lattice. There are other lattice systems that occur in nature, but they are outside the scope of this text. There are three types of cubic unit cells within the cubic system. Before we get into the specifics of the crystal lattice structure, let's review some basic geometry.

Basic Geometry Review:

Radius: the line segment that connects the center of the sphere to the edge, represented by r

Volume of a sphere = $\frac{4}{3}\pi r^3$

Volume of a cube = a³, where a is the edge length of the cube

Sphere

Cube

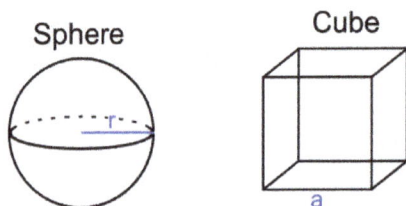

Figure 3. Basic geometry review

Simple Cubic

In a simple cubic unit cell, Figure 4A, the centers of 8 atoms define the corners of the cube. Each of the 8 corners is 1/8 of an atom. The entire unit cell contains 8 x 1/8 atoms = 1 atom. The edge length of each cell is 2 x radius of the atom, or 2r. The volume of the simple cubic unit cell would be the edge length cubed which is (2r)³.

Another way to think about simple cubic is placing each sphere or atom on top of each other. Each layer of the atoms is identical and is stacked on each atom of that layer.

A. Simple Cubic

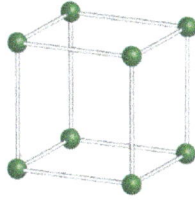

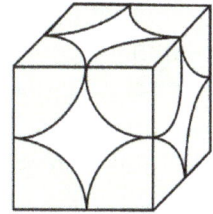

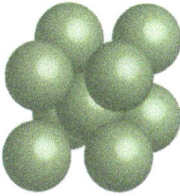

B. Body Centered Cubic

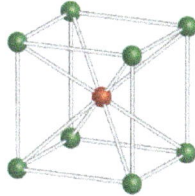

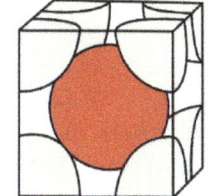

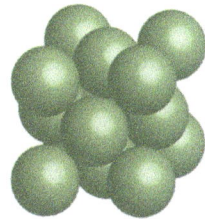

C. Face Centered Cubic

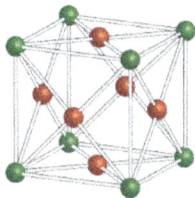

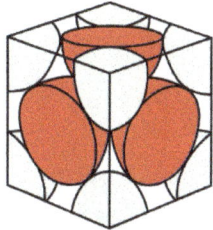

Figure 4. The three cubic unit cells. A. simple; B. body-centered and C. face centered cubic. Left column: space filling view of these cubic arrangements. Middle column: cubic arrangement of atoms in expanded view. Right column: the atoms in the actual unit cells.

Body Centered Cubic

In body centered cubic, BCC, the layers of atoms differ. The atoms or spheres are placed in the diamond shaped space between the 4 atoms of the first layer (Figure 4B left side), then again on the alternating layers. The layers of body centered cubic alternate between two layers. The resulting unit cell has an entire atom in the middle with each corner as the center of another atom as is the case with the simple cubic. The entire unit cell contains eight 1/8th corner atom pieces along with the atom in the center, which results in the BCC unit cell containing 2 atoms.

To determine the volume of the cube we need to figure out the edge length. We need to look at the dimensions of a cube. We see that the longest diagonal in the cube going from one corner to the opposite core, shown in Figure 5 in red, is equal to the side length multiplied by the square root of 3.

This was derived by using the Pythagorean theorem for the triangle formed by side a, the diagonal of a face of the cube, d_1 and the diagonal of the cube as the hypotenuse, d.

$$a^2 + d_1{}^2 = d^2$$

Before we can solve for d, we first need to solve for d_1. d_1 is the diagonal of the face of the cube, we can use the Pythagorean theorem to solve for its length.

$$a^2 + a^2 = d_1{}^2$$

$$2a^2 = d_1{}^2$$

$$d_1 = \sqrt{(2a^2)}$$

$$d_1 = a\sqrt{2}$$

Now to find d, we use the Pythagorean theorem and substitute in our value for d_1 to the formula.

$$a^2 + d_1{}^2 = d^2$$

$$a^2 + (a\sqrt{2})^2 = d^2$$

$$a^2 + 2a^2 = d^2$$

$$3a^2 = d^2$$

$$d = \sqrt{(3a^2)}$$

$$d = a\sqrt{3}$$

In looking at Figure 5, we see that the d is equal to 4 times the radius or 4r, based on the packing shown in Figure 4b.

$$4r = a\sqrt{3}$$

$$r = \frac{a\sqrt{3}}{4} \text{ or } a = \frac{4r}{\sqrt{3}}$$

In order to calculate the volume of the cube, we need to cube the edge length, which we now know is $a = \frac{4r}{\sqrt{3}}$.

So, the volume of a BCC cell is $\left(\frac{4r}{\sqrt{3}}\right)^3$.

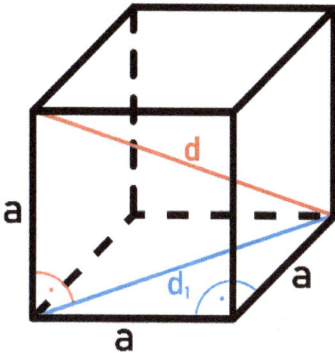

Figure 5. The diagonal measurements within a cube.

Face Centered Cubic

In face centered cubic, FCC, the layers of atoms are packed most efficiently. They are packed in a cubic closest packing pattern in which three different layers are repeated in an *abcabc...* pattern. The resulting unit cell has each corner is occupied by 1/8 of an atom and each face contains half of an atom (Figure 4C). Each FCC unit cell contains 4 atoms, as each corner is 1/8 of an atom and each of the 6 faces of the cube contains ½ an atom (8 x 1/8 + 6 x ½).

For the volume of the FCC, we see in Figure 4C, that the diagonal along a face of the cube, d_1, is 4r. We solved for d_1 in the BCC calculation above, so we know that $d_1 = a\sqrt{2} = 4r$, after rearranging this equation, we find that the edge length of the FCC is $\frac{4r}{\sqrt{2}}$ and the volume of the FCC would be the edge length cubed which gives us $\left(\frac{4r}{\sqrt{2}}\right)^3$.

Table 4. Summary of Lattice Cubic Cells

	SIMPLE CUBIC	BODY CENTERED CUBIC	FACE CENTERED CUBIC
# ATOMS IN EACH CELL	1	2	4
EDGE LENGTH OF CUBE	2r	$\dfrac{4r}{\sqrt{3}}$	$\dfrac{4r}{\sqrt{2}}$
VOLUME OF CUBE	$(2r)^3$	$\left(\dfrac{4r}{\sqrt{3}}\right)^3$	$\left(\dfrac{4r}{\sqrt{2}}\right)^3$

Packing efficiency derivations

The amount of occupied space or percent density indicates the percentage of packing efficiency of each of the unit cells. We can determine the packing efficiency of each cubic cell by determining the ratio of the volume of the atoms to the volume of the cell in the following equation:

$$\% \, density = \frac{Volume_{atoms}}{Volume_{\;cube}} \times 100\%$$

Simple Cubic Packing Efficiency:

To determine this percentage, we need to take the part occupied and divide it by the entire cube, then multiply by 100% (see percentage calculation section)

The part that is occupied is the volume of the atoms in the unit cell, and the volume of the entire cube is the entirety of the sample.

$$\% \, density = \frac{Volume_{atoms}}{Volume_{\;cube}} \times 100\%$$

To determine the volume of the atoms, we take the number of atoms in the unit cell and multiply it by the volume of a sphere. The Volume of the cube is the volume that we calculated in the previous section, which is summarized in Table 4.

$$\% \text{ } density = \frac{1 \times \frac{4}{3}\pi r^3}{(2r)^3} \times 100 = \frac{\frac{4}{3}\pi r^3}{8r^3} \times 100 = \frac{\pi}{6} \times 100 = 52\%$$

For a simple cubic unit cell, 52% of the density is occupied by the atoms.

Body Centered Cubic Packing Efficiency:

To solve this, we will need to figure out the volume of the atoms and the volume of the BCC cell and plug those values into the % density equation below.

$$\% \text{ } density = \frac{Volume_{atoms}}{Volume_{cube}} \times 100\%$$

Based on Table 4, we see that the BCC has 2 atoms per unit cell and has a volume of $\left(\frac{4r}{\sqrt{3}}\right)^3$. Using the volume of a sphere, we can determine the volume of the two atoms within the BCC cell. The volume of the cube will be the volume of the BCC we derived in the previous section.

$$\% \text{ } density = \frac{2 \times \frac{4}{3}\pi r^3}{\left(\frac{4r}{\sqrt{3}}\right)^3} \times 100 = \frac{2 \times 4 \times \frac{1}{3}\pi r^3}{\frac{4 \times 4 \times 4 \times r^3}{3\sqrt{3}}} \times 100$$

$$= \frac{2\pi}{\frac{16}{\sqrt{3}}} \times 100 = \frac{\pi\sqrt{3}}{8} \times 100 = 68\%$$

When we cube the denominator we get 4^3, which is the same as writing $4 \times 4 \times 4$, and we also cube $(\sqrt{3})^3$, which is the same as $\sqrt{3} \times \sqrt{3} \times \sqrt{3}$, which simplifies to $3\sqrt{3}$. After canceling and multiplying the numerator and denominator by $\sqrt{3}$, we are left with $\frac{\pi\sqrt{3}}{8} \times 100\%$, which is roughly 68%. The volume of the atoms within the BCC occupy approximately 68% of the total volume, which is more efficient than the simple cubic structure.

Face Centered Cubic Packing Efficiency:

Using Table 4, we see that the face centered cubic cell contains 4 atoms and has a volume of $\left(\frac{4r}{\sqrt{2}}\right)^3$. The density of the atoms within the cell would then be:

$$\% \: density = \frac{Volume_{atoms}}{Volume_{cube}} \times 100\%$$

Where the volume is 4 times the volume of a sphere divided by the volume of the FCC unit cell.

$$\% \: density = \frac{4 \times \frac{4}{3}\pi r^3}{\left(\frac{4r}{\sqrt{2}}\right)^3} \times 100 = \frac{\frac{16}{3}\pi r^3}{\frac{64 \times r^3}{2\sqrt{2}}} \times 100 = \frac{16\pi 2\sqrt{2}}{3 \times 64} \times 100$$

$$= \frac{\pi\sqrt{2}}{6} \times 100 = 74\%$$

The face centered cubic packing has the highest efficiency. There is no way to pack spheres of equal size more efficiently.

Example Problems:

Other problems you may come across dealing with unit cell calculation may include giving you the radius of an element and the type of crystal lattice that forms and asking you to determine the density of that particular element. Or the question could give you the lattice information and the density and ask you to find the radius.

Both of these types of questions will involve the same equation:

$$density = \frac{mass}{volume}$$

The mass can be found on the periodic table and the volume will based on the type of lattice. Like with any chemistry question, it is important to look at what information you have and what you are being asked to figure out so that you can set up the equation correctly.

Consider the following problem: Gallium forms a simple cubic crystal lattice. If the radius of an atom of gallium is 141 pm, calculate the density of gallium.

Here we know the lattice and in turn the volume equation and we are also given the radius. We are being asked to determine the density of gallium. A problem that may arise within the calculation is with the units. The radius is given in terms of pm but the density is typically given in g/cm³. So unit conversions are necessary.

Let's first determine the mass of gallium. If we look at the periodic table, we can see that the mass is 69.72 amu. Since we know our density will be given in terms of g/cm³, we need to convert amu to grams.

$$mass = 69.7 \ amu \ \times \frac{1 \ g}{6.02 \times 10^{23} \ amu} = 1.16 \times 10^{-22} g$$

Next, let's look at the volume. Based on the question, we know that gallium forms a simple cubic crystal lattice which we know has a volume of (2r)³. The radius is given in terms of picometers, but our answer needs to be in terms of centimeters so additional unit conversions are necessary.

$$radius = 141 pm \ \times \frac{10^{-12} \ m}{1 \ pm} \times \frac{100 cm}{1 m} = 1.41 \times 10^{-8} cm$$

$$volume = (2 \times 1.41 \times 10^{-8} cm)^3 = (2.82 \times 10^{-8} \ cm)^3$$
$$= 2.242 \times 10^{-23} cm^3$$

Now let's put both the mass and volume in the following equation:

$$density = \frac{mass}{volume} = \frac{1.16 \times 10^{-22} g}{2.242 \times 10^{-23} cm^3} = 5.16 \ g/cm^3$$

We find that the density of Gallium is 5.16 g/cm³.

Here is another crystal structure example problem. Gold has a density of 19.32 g/cm³. X-ray crystallography reveals that the edge length of its unit cell is 4.07Å. Determine the type of cubic packing for gold. [Hint: 1Å = 1 x 10⁻¹⁰m]

In this problem we are given the element, the density and the edge length of the unit cell which we can use to determine the volume. We are being asked to determine the type of packing which we can determine by the number of gold atoms present in the unit cell.

We know that:

$$density = \frac{mass}{volume}$$

We are given the density and a way to find the volume by cubing the edge length. So if we rearrange the equation and solve for mass we get:

$$mass = density \times volume$$

Since the density is given in terms of cm, we need to convert the edge length, Å, to cm.

$$4.07\text{Å} \times \frac{1 \times 10^{-10}m}{1\text{Å}} \times \frac{100 \ cm}{1 \ m} = 4.07 \times 10^{-8} \ cm$$

If we plug in the values for density and volume, we get

$$mass = \frac{19.32 \ g}{cm^3} \times (4.07 \times 10^{-8} \ cm)^3 = 1.30 \times 10^{-21} g$$

We also know that the mass is equal to the number of atoms in the unit cell multiplied by the atomic weight dived by Avogadro's number to get the mass in the units of grams.

$$1.30 \times 10^{-21} g = mass$$
$$= \#atoms \times atomic \ weight \times \frac{1g}{6.02 \times 10^{23} \ amu}$$

If we plug in the atomic weight of gold which is found on the periodic table, we find:

$$1.30 \times 10^{-21} g = \#atoms \times 197.0 \, amu \times \frac{1g}{6.02 \times 10^{23} \, amu}$$

$$= \#atoms \times 3.27 \times 10^{-22} g$$

Which we can simplify further and solve for the # of atoms:

$$\#atoms = \frac{1.30 \times 10^{-21} g}{3.27 \times 10^{-22} g} = 3.98 \, atoms$$

We find that the number of atoms rounds to 4 and based on Table 4, we know that the unit cell that contains 4 atoms is the face centered cubic. Gold has a face centered cubic lattice structure.

Consider this last example problem: Barium, an alkaline earth metal, has a body centered cubic unit cell and a density of 3.62 g/cm³. What is the atomic radius of barium?

In this problem we are given the packing and the density and want to find out the radius. So we will need to rearrange the density equation and solve it in terms of r, radius to get our answer.

$$density = \frac{mass}{volume}$$

Which can be rearranged to:

$$volume = \frac{mass}{density}$$

As defined by the problem, we know that barium has a body centered cubic packing and based on Table 4, we know that the volume of BCC is $\left(\frac{4r}{\sqrt{3}}\right)^3$. By replacing the volume with the volume of a BCC we can now solve for the radius.

$$volume = \left(\frac{4r}{\sqrt{3}}\right)^3 = \frac{mass}{density}$$

If we cube the volume term, we get $\frac{4^3 \times r^3}{\sqrt{3} \times \sqrt{3} \times \sqrt{3}}$, which simplifies to $\frac{64r^3}{3\sqrt{3}}$. Plugging values into the simplified volume equation makes it easier to solve for r.

$$\frac{64\,r^3}{3\sqrt{3}} = \frac{mass}{density}$$

We can rearrange and get the radius on one side and the other terms on the other.

$$r^3 = \frac{mass \times 3\sqrt{3}}{64 \times density}$$

So

$$r = \sqrt[3]{\frac{mass \times 3\sqrt{3}}{64 \times density}}$$

Now that we know what r equals, let's put the mass and density of barium in the correct units.

$$mass = 2 \times 137.3\ amu \times \frac{1\,g}{6.02 \times 10^{23}\,amu} = 4.56 \times 10^{-22}\,g$$

Recall that the unit cell of the body centered cubic has two atoms, so we need to take the mass of barium and multiply it by two to account for both atoms in the unit cell.

Based on the problem, the density is given in g/cm³, so we don't need to do any conversions this time.

$$r = \sqrt[3]{\frac{(4.56 \times 10^{-22}\,g) \times 3\sqrt{3}}{64 \times 3.62\,\dfrac{g}{cm^3}}}$$

$$r = \sqrt[3]{\frac{2.37 \times 10^{-21}\,g}{231.7\,\dfrac{g}{cm^3}}}$$

$$r = \sqrt[3]{1.02 \times 10^{-23}\,cm^3}$$

$$r = 2.17 \times 10^{-8}\ cm$$

The radius for barium is 2.17 x 10⁻⁸ cm.

Additional Crystal Structure Problems (9.1)

Iridium crystallizes in a face-centered cubic unit cell that has an edge length of 3.833 Å. What is the density of iridium metal and what is the atomic radius of an iridium atom?

Calcium crystallizes with a body-centered cubic structure. How many Ca atoms are contained in each unit cell? If the radius of calcium is 1.97 Å, what is the length of the unit cell? What is the density of Ca metal?

Graphing

Within a lab, graphs are often used to show the relationship between two properties. The graphs used in chemistry will have one property plotted on the y-axis and the other property plotted on the x-axis. The majority of graphs you will encounter in the general chemistry will be a linear function that follows the slope intercept equation:

$$y = mx + b$$

The slope of the graph is m. It represents the ratio of the change in the rise (y) to the change in the run (x). Since the line is linear, the slope is the same between any two points along the line.

The slope, m, is defined as the rise over the run.

$$m = \frac{rise}{run} = \frac{change\ in\ y}{change\ in\ x} = \frac{y_2 - y_1}{x_2 - x_1}$$

The b term is called the y-intercept and is the value of y when x is zero.

The data that gets plotted on the x and y axis depends on what graphical information you want to show.

For example, a common lab in general chemistry is a density lab, where you are tasked with measuring the mass of copper objects of known volume, then determining the density by taking the mass and dividing it by the volume.

$$density = \frac{mass}{volume}$$

If you are asked to determine the density of the copper, you could find the density of each object by dividing the mass by the volume then take the average of all the measurements. If there are a few data points, this approach is feasible. However, if you had over 100 measurements this approach would be cumbersome. A more streamlined approach would be to graph the data so that the slope is representative of the density (mass/volume).

Table 5. Example Experimental Data of the mass measured for copper objects of known volume.

Object Number	Volume (cm³)	Mass (g)
1	0.5	4.50
2	1.0	8.75
3	1.5	13.4
4	2.5	22.3
5	3.5	31.4
6	5.8	51.5

The slope is the rise over run and for it to be the density, it would need to be:

$$m = \frac{rise\ (y)}{run\ (x)} = \frac{mass}{volume}$$

If the mass is plotted on the y-axis and the volume is plotted on the x-axis, the slope, m, will be equal to the density. If we plot the data in Table 5, we obtain the following graph.

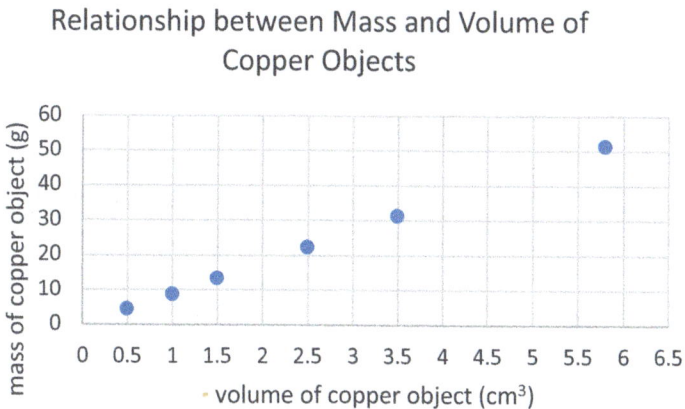

Figure 6. Data from Table 5 with mass plotted on y- axis and volume plotted on x-axis

To determine the density, you can have the plotting software determine the slope or you can take two data points and determine the slope that way.

$$m = \frac{rise}{run} = \frac{change\ in\ y}{change\ in\ x} = \frac{y_2 - y_1}{x_2 - x_1} = \frac{(51.5 - 22.3)g}{(5.8 - 2.5)cm^3} = \frac{29.2g}{3.3cm^3}$$

$$= 8.8484\frac{g}{cm^3} = 8.8\frac{g}{cm^3}$$

Upon finding the slope from the data in Table 5, the experimental density of the copper objects is found to be 8.8 g/cm³.

Another example where graphing is useful would be with labs involving the ideal gas law. The ideal gas law is:

$$PV = nRT$$

The variable P is pressure, V is volume, n is the number of moles, R is the gas constant, and T is temperature. Assuming that the amount of gas, n, is constant, one can fix one of the remaining parameters (P, V or T) and establish a relationship with the other two.

If the pressure, P, remains constant, then $V = \frac{nR}{P}T$, where the $\frac{nR}{P}$ term is a constant value.

If the volume, V, remains constant, then $P = \frac{nR}{V}T$, where the $\frac{nR}{V}$ term is a constant.

If the temperature, T, remains constant, then $P = \frac{nRT}{V}$, where the nRT term is a constant.

Consider you are given the following graph of pressure vs. temperature of an experiment where the volume remained at a constant of 0.50L, Figure 7. With the volume remaining constant and the relationship between pressure and temperature shown in graph form, we can determine the number of moles present in the vessel. The y-intercept formula has been determined to be:

y= 0.0033x

Pressure vs. Temperature

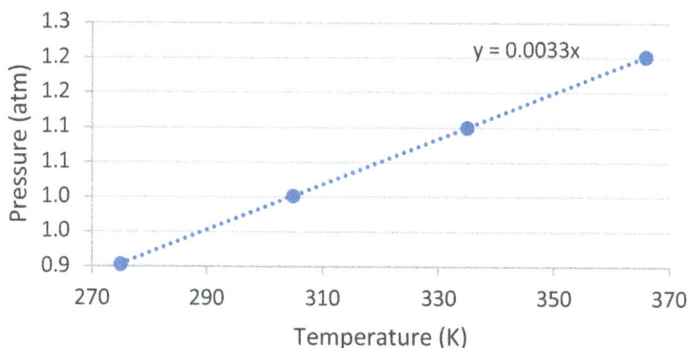

Figure 7. Experimental Data of Pressure vs. Temperature of a gas where the volume remained at a constant 0.5L.

Since the volume is constant, we know that the relationship between the Pressure and Temperature is represented by P = (constant) T. This is the same format as $y = mx+b$. In this case, the constant is the slope, m, and we know that with a fixed volume, the ideal law can be written with a constant of $\frac{nR}{V}$. The slope for this example is 0.0033 which is equal to nR/V.

$$0.0033 = \frac{nR}{V}$$

R is the universal gas constant which is $0.0821\frac{L \cdot atm}{mol \cdot K}$, and V in this example is 0.5L, so if we rearrange the equation and solve for n, we get:

$$\frac{0.0033\ V}{R} = n$$

Once we plug in the known values for the variables, we get:

$$\frac{0.0033\ \frac{atm}{K}\ (0.5\ L)}{0.0821\ \frac{L \cdot atm}{mol \cdot K}} = n$$

We find that

$$n = 0.020\ mol$$

Consider the following data, where the rate of decomposition of H_2O_2 in an aqueous solution decreases as the concentration of H_2O_2 decreases.

Table 6. Hypothetical data for the decomposition of H_2O_2

Time (hours)	$[H_2O_2](M)$	$\ln[H_2O_2]$
0.00	2.000	0.6931
3.00	1.000	0.0000
6.00	0.500	-0.6931
9.00	0.250	-1.3863
12.00	0.125	-2.0794
15.00	0.0625	-2.7726

If you were asked to show that the reaction represented a first-order rate law and asked to find the rate constant, k, graphing the data would be the quickest way to confirm the first-order rate law and find k.

Based on the definition of the rate law, we know that for a first-order reaction,

$$[A]_t = [A]_0 e^{-kt}$$

where $[A]_t$ is the concentration of A at a given time, t, $[A]_0$ is the initial concentration, and k is the first order-rate constant. To make this look like an equation we are more familiar with for graphing, we can rearrange the format of the equation. If we take the natural log (ln) of both sides of the above equation, we find:

$$\ln([A]_t) = \ln([A]_0 e^{-kt})$$

When you take a natural log of a product, the result can be rewritten as a sum of the two natural logs. So the above equation simplifies to:

$$\ln([A]_t) = \ln([A]_0) + \ln(e^{-kt})$$

When you take the natural log of an exponent, it can be reduced to just the exponent as the natural log, ln, is a log with a base of e (approximately 2.718). The above equation is the simplified to:

$$\ln([A]_t) = \ln([A]_0) - kt$$

66

Which if we rearrange the products, we find:

$$\ln([A]_t) = -kt + \ln([A]_0)$$

This now looks similar to the slope intercept form of:

$$y = mx + b$$

Now, y is $\ln[A]_t$, the slope is -k, x is t, and b is $\ln[A]_0$. So if we plot the data for $\ln[A]_t$ on the y axis and data for time on the x axis, we should be able to figure out the rate constant k, as it will be the slope of the line (Table 6).

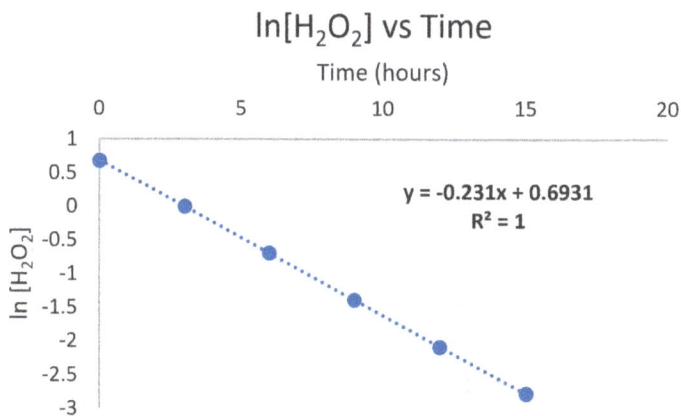

Figure 8. The graph of ln[H₂O₂] vs time.

Figure 8 shows the linear relationship between the $\ln[H_2O_2]$ and time which is what is expected for a first order rate reaction. Based on this, we can confirm that the decomposition of H_2O_2 is a first order reaction.

To find k, we can look to the value of the slope as the slope is equal to -k. The slope of this data is -0.231 so we can conclude that k is 0.231 h^{-1}. Alternatively, we can find k by computing the slope as the change in the rise divided by the change in the run.

$$m = \frac{rise}{run} = \frac{change\ in\ y}{change\ in\ x} = \frac{y_2 - y_1}{x_2 - x_1} = \frac{(0.6931 - (-0.6931))}{(0 - 6)hours}$$

$$= \frac{1.386}{-6\ hours} = -0.2310\frac{1}{h}$$

Practice Problem Solutions

1.1

$$17 + 21 = 38 \qquad\qquad 5 * 3 = 15$$

$$29 - 13 = 16 \qquad\qquad 9 \div 3 = 3$$

$$8 \times 7 = 56 \qquad\qquad \frac{32}{8} = 4$$

$$\frac{21}{7} = 3 \qquad\qquad (32)(4) = 128$$

1.2

$$4 + -7 = -3$$

$$-5 + -5 = -10$$

$$31 - (-27) = 58$$

$$-6 \div -2 = 3$$

$$5 \times -6 = -30$$

$$-7 \times (-3) = 21$$

1.3

$$(11 + -9) \times 4 + 5 - 21 \div 7 =$$

$$2 \times 4 + 5 - 21 \div 7 =$$

$$8 + 5 - 21 \div 7 =$$

$$8 + 5 - 3 = 10$$

$$(37 - 25) \div 6 + 7 =$$

$$12 \div 6 + 7 =$$

$$2 + 7 = 9$$

$$3^2 - 16 \div 4 + 32 =$$

$$9 - 16 \div 4 + 32 =$$

$$9 - 4 + 32 = 37$$

1.4 For additional practice, solve the ideal gas law for P, V and n.

$PV = nRT$ to isolate P, divide both sides of the equation by V to obtain:

$$P = \frac{nRT}{V}$$

$PV = nRT$ to isolate V, divide both sides of the equation by P to obtain:

$$V = \frac{nRT}{P}$$

$PV = nRT$ to isolate n, divide both sides of the equation by RT to obtain:

$$\frac{PV}{RT} = n$$

1.5 $M_1V_1 = M_2V_2$, to solve for V_2 we need to divide both sides of the equation by M_2 to obtain:

$$\frac{M_1 V_1}{M_2} = V_2$$

$\Delta G = \Delta H - T\Delta S$, to solve for ΔH, we need to add $T\Delta S$ to both sides of the equation

$$\Delta G + T\Delta S = \Delta H$$

$\Delta G = \Delta H - T\Delta S$, to solve for ΔS, we first need to subtract ΔH to both sides of the equation

$$\Delta G - \Delta H = -T\Delta S$$

Next we need to divide both sides of the equation by -T to obtain:

$$\frac{-\Delta G + \Delta H}{T} = \Delta S$$

$\Delta G = \Delta H - T\Delta S$, to solve for ΔS, we first need to subtract ΔH to both sides of the equation

$$\Delta G - \Delta H = -T\Delta S$$

Next, we need to divide both sides of the equation by -ΔS to obtain:

$$\frac{-\Delta G + \Delta H}{\Delta S} = T$$

2.1

0.0040 L has 2 sig figs, 0.00$\underline{40}$

0.125607 mg has 6 sig figs, 0.$\underline{125607}$

53,608 cm³ has 5 sig figs, $\underline{53,608}$

0.00000000001215 km has 4 sig figs, 0.0000000000$\underline{1215}$

83,800. s has 5 sig figs, $\underline{83,800}$.

156890000 mL has 5 sig figs, $\underline{156890}$000

2.2

What is the density of an unknown metal that has a mass of 13.1g and a volume of 262.5 cm³?

$$density = \frac{mass}{volume} = \frac{13.1g}{262.5cm^3} = 0.049904761 \frac{g}{cm^3}$$

We need to round, so that the answer has 3 sig figs as that is the number of sig figs in the measurement with the least amount of certainty. So 0.049904761 rounds to $0.0499 \frac{g}{cm^3}$.

Calculate the following:

$$\frac{63.2g + 78g + 636.25\ g}{7.2\ g}$$

$$\frac{777.45\ g}{7.2\ g} = 107.979167$$

This needs to be rounded to reflect the measurement with the lowest degree of certainty which is 78g. So the answer would be 108.

$$\frac{16.7451\ cm - 3.85\ cm}{2.356\ cm}$$

$$\frac{12.8951\ cm}{2.356\ cm} = 5.47330221$$

This value needs to be rounded to reflect the measurement with the lowest number of decimal places which in this example is 3.85cm. So, the answer should have two decimal places and thus will be rounded to 5.47.

3.1

63.809	→ 6.3809×10^1
0.0000325068	→ 3.25068×10^{-5}
631,587,598	→ 6.31587598×10^8
0.0000000002598	→ 2.598×10^{-10}
1.235854×10^{-7}	→ 0.0000001235854
3.2587×10^9	→ 3,258,700,000
2.3568×10^6	→ 2,356,800
$5.6874298 \times 10^{-14}$	→ 0.000000000000056874298

$$\frac{(5.2 \times 10^{23}) \times (3.2 \times 10^{-16})}{6.1 \times 10^{-8}} =$$

$$\frac{16.64 \times 10^{(23+ -16)}}{6.1 \times 10^{-8}} =$$

$$\frac{16.64 \times 10^7}{6.1 \times 10^{-8}} =$$

$$\frac{1.664 \times 10^8}{6.1 \times 10^{-8}} =$$

$$0.272787 \times 10^{[8-(-8)]} =$$

$$0.272787 \times 10^{16} =$$

$$2.72787 \times 10^{15} =$$

Remember we need to round so that our answer only has 2 significant digits

$$2.7 \times 10^{15}$$

$$\frac{2.3 \times 10^{-6}}{(9.63 \times 10^{-3})(3.11 \times 10^{-5})} =$$

$$\frac{2.3 \times 10^{-6}}{29.9493 \times 10^{[-3+(-5)]}} =$$

$$\frac{2.3 \times 10^{-6}}{29.9493 \times 10^{-8}} =$$

$$\frac{2.3 \times 10^{-6}}{2.99493 \times 10^{-7}} =$$

$$0.767964527 \times 10^{[-6-(-7)]} =$$

$$0.767964527 \times 10^1 =$$

$$7.67964527 =$$

$$7.7$$

$$\frac{4 * (3.2 \times 10^{-8})}{9.1 \times 10^{-10}} =$$

$$\frac{12.8 \times 10^{-8}}{9.1 \times 10^{-10}} =$$

$$\frac{1.28 \times 10^{-7}}{9.1 \times 10^{-10}} =$$

$$0.14066 \times 10^{[-7-(-10)]} =$$

$$0.14066 \times 10^3 =$$

$$1.4066 \times 10^2 =$$

The question has one number with 1 sig fig so the answer needs to have one sig fig so in this case our answer is rounded to 100 or 1 x 10^2.

4.1

What is the pH of 0.001 M H_2SO_4?

0.001 M H_2SO_4 → 1 x 10^{-3} M H_2SO_4

$[H^+] = 10^{-pH}$

$10^{-3} = 10^{-pH}$

pH = 3

Calculate the pH of 0.00001 M NaOH

0.00001 M NaOH → 1 x 10^{-5} M NaOH

$10^{-pOH} = [OH^-] = 10^{-5}$

pOH = 5

pH + pOH = 14 → 14-pOH = pH

14 - 5 = 9

4.2

What is the pH of a solution containing 0.25 mol of NH_4Cl and 2.5 mol of NH_3? The K_b of NH_3 is 1.8×10^{-5}.

We need to first figure out the pKa with the information given by finding the pK_b then find the pK_a.

$pK_b = -\log K_b = -\log (1.8 \times 10^{-5}) = 4.74473$

$pK_a = 14 - pK_b = 14 - 4.74473 = 9.25527 = 9.3$

Since this is a buffer system, we can use the Hendersen-Hasselbach equation:

$$pH = pKa + \log \left(\frac{[Base]}{[Acid]}\right)$$

$$pH = 9.3 + \log \left(\frac{2.5\ mol}{0.25\ mol}\right)$$

$$pH = 9.3 + 1 = 10.3$$

Calculate the pH of a solution containing 25g HF and 31g NaF in 750mL of solution. The K_a of HF is 7.2×10^{-4}.

Since HF and NaF are conjugate pairs, this is a buffer system problem in which we can use the Hendersen-Hasselbalch formula. To start we need to determine the number of moles of both NaF and HF and the pK_a of HF. Once these have been determined, the values can be plugged into the H-H equation to solve for the pH.

$$31g\ NaF \times \frac{1\ mol\ NaF}{42g\ NaF} = 0.738\ mol\ NaF$$

$$25g\ HF \times \frac{1\ mol\ HF}{20g\ HF} = 1.25\ mol\ HF$$

$pK_a = -\log K_a = pK_a = -\log (7.2 \times 10^{-4}) = 3.1427$

$pH = pK_a + \log (B/A) = 3.1427 + \log (0.738\ mol\ NaF/1.25\ mol\ HF)$

$pH = 2.913 = 2.9$

5.1

How many seconds are in 1 nanosecond? in decisecond?

1 nanosecond = 1/1000000000 seconds or 1 nanosecond is 0.000000001 seconds.

1 decisecond = 1/10 seconds or 1 decisecond is 0.1 seconds

How many picograms are in 1 gram? in 1 kilogram?

$$1\,g \times \frac{1000000000000\,pg}{1g} = 1000000000000\,pg$$

1 gram = 1000000000000 picogram

1000 gram = 1 kilogram or 1 gram = 0.0001 kilograms

$$1\,kg \times \frac{1000\,g}{1\,kg} \times \frac{1000000000000\,pg}{1\,g} = 1000000000000000\,pg$$

1 kilogram = 1000000000000000 picogram

How many milliliters are in 1 liter? In 1 megaliter?

$$1\,l \times \frac{1000\,ml}{1\,l} = 1000\,ml$$

There are 1000 mL in 1 liter.

$$1\,Ml \times \frac{1000000\,l}{1\,Ml} \times \frac{1000\,ml}{1\,l} = 1000000000\,ml$$

There are 1000000000 mL in 1 megaliter.

6.1

What is the price of a piece of 325 cm long copper wire that costs $0.90/ft? [hint: 1in = 2.54 cm]

$$325\,cm \times \frac{1\,in}{2.54\,cm} \times \frac{1\,ft}{12\,in} = 10.66\,ft$$

$$10.66 \ ft \ \times \ \frac{\$0.90}{ft} = \$9.60$$

The radius of a copper atom is 128pm. What is its radius in nanometers (nm)?

1 pm = 10^{-12} m   1 nm = 10^{-9} m

$$128 \ pm \ \times \ \frac{10^{-12} \ m}{1 \ pm} \ \times \ \frac{1 \ nm}{10^{-9} \ m} = 0.128 \ nm$$

How many grams are in 0.428 moles of sodium bicarbonate, $NaHCO_3$?

$NaHCO_3$ MW → Na 23 g/mol + H 1 g/mol + C 12 g/mol + 3O (16 g/mol x 3) 48 g/mol

$NaHCO_3$ MW = 23g/mol + 1g/mol + 12 g/mol + 48 g/mol = 84 g/mol

$$0.428 \ mol \ \times \ \frac{84 \ g}{mol} = 35.952 \ g = 36.0 \ g$$

7.1

Figure out how much H_2O is produced from the combustion of 1.0g of butane.

The reaction for the combustion of butane is as follows:

$$4 \ C_4H_{10} + 13 \ O_2 \rightarrow 8 \ CO_2 + 10 \ H_2O$$

To determine the amount of H_2O produced we need to convert 1.0g of C_4H_{10} to moles, then convert moles C_4H_{10} to moles of H_2O using their coefficients then convert moles H_2O to grams H_2O using molar mass of H_2O.

Game plan: g C_4H_{10} → mol C_4H_{10} → mol H_2O → g H_2O

$$1.0 \ g \ C_4H_{10} \times \frac{1 \ mol \ C_4H_{10}}{58.0g \ C_4H_{10}} \times \frac{10 \ mol \ H_2O}{4 \ mol \ C_4H_{10}} \times \frac{18.0g \ H_2O}{1 \ mol \ H_2O} = 0.78g \ H_2O$$

Sodium bicarbonate and citric acid react together in an acid base reaction as follows:

$$3NaHCO_3(aq) + H_3C_6H_5O_7(aq) \rightarrow 3CO_2(g) + 3H_2O(l) + Na_3C_6H_5O_7(aq)$$

If 1.21g of sodium bicarbonate and excess citric acid react, how many g of CO_2 form?

Game plan: g $NaHCO_3 \rightarrow$ mol $NaHCO_3 \rightarrow$ mol $CO_2 \rightarrow$ g CO_2

$$1.21\ g\ NaHCO_3 \times \frac{1\ mol\ NaHCO_3}{84\ g\ NaHCO_3} \times \frac{3\ mol\ CO_2}{3\ mol\ NaHCO_3} \times \frac{44\ g\ CO_2}{1\ mol\ CO_2}$$
$$= 0.633\ g\ CO_2$$

7.2

How many grams of solid aluminum sulfide can be prepared by the reaction of 5.0 g of aluminum and 10.0 g sulfur?

$$2Al + 3S \rightarrow Al_2S_3$$

Game plan: g Al $\rightarrow$ mol Al $\rightarrow$ mol $Al_2S_3 \rightarrow$ g Al_2S_3 and g S $\rightarrow$ mol S $\rightarrow$ mol $Al_2S_3 \rightarrow$ g Al_2S_3 then determine which amount is the smallest and that is the answer

$$5.0\ g\ Al \times \frac{1\ mol\ Al}{27\ g\ Al} \times \frac{1\ mol\ Al_2S_3}{2\ mol\ Al} \times \frac{150.2\ g\ Al_2S_3}{1\ mol} = 13.9\ g\ Al_2S_3$$

$$10.0\ g\ S \times \frac{1\ mol\ S}{32\ g\ S} \times \frac{1\ mol\ Al_2S_3}{3\ mol\ Sl} \times \frac{150.2\ g\ Al_2S_3}{1\ mol} = 15.6\ g\ Al_2S_3$$

Aluminum is the limiting reagent so only 13.9 g Al_2S_3 can be produced.

Sodium hydroxide reacts with carbon dioxide as follows:

$$2NaOH(s) + CO_2(g) \rightarrow Na_2CO_3(s) + H_2O(l)$$

If 1.75 mol NaOH and 2.50 mol CO_2 are allowed to react, what is the limiting reactant? How many moles of water will be produced?

Game plan: mol NaOH → mol H_2O and mol CO_2→ mol H_2O then determine which reactant produced the smallest amount of H2O and that is your limiting reactant and the amount of water that will be produced

$$1.75 \; mol \; NaOH \; \times \frac{1 \; mol \; H_2O}{2 \; mol \; NaOH} = 0.875 \; mol \; H_2O$$

$$2.5 \; mol \; CO_2 \times \frac{1 \; mol \; H_2O}{1 \; mol \; CO_2} = 2.5 \; mol \; H_2O$$

NaOH is the limiting reagent, only 0.875 mol H_2O will be produced.

How much molten iron is produced from the reaction of 35.0g FeO with 25.0g Al? Given the balanced equation:

$$3FeO(l) + 2Al(l) \rightarrow 3Fe(l) + Al_2O_3(s)$$

Game plan: g FeO → mol FeO → mol Fe → g Fe and g Al → mol Al → mol Fe → g Fe The amount with the least amount of Fe is the amount that is produced

$$35.0 \; g \; FeO \; \times \frac{1 \; mol \; FeO}{71.8 \; g \; FeO} \times \frac{3 \; mol \; Fe}{3 \; mol \; FeO} \times \frac{55.8 \; g \; Fe}{1 \; mol \; Fe} = 27.2 \; g \; Fe$$

$$25.0 \; g \; Al \; \times \frac{1 \; mol \; Al}{27.0 \; g \; Al} \times \frac{3 \; mol \; Fe}{2 \; mol \; Al} \times \frac{55.8 \; g \; Fe}{1 \; mol \; Fe} = 77.5 \; g \; Fe$$

FeO is the limiting reagent so only 27.2 g Fe will be produced.

8.1 What is the percent hydrogen and the percent oxygen of $C_6H_{12}O_6$.

$$\frac{(mass \; of \; Hydrogen)(\# \; of \; Hydrogens)}{mass \; of \; C_6H_{12}O_6} \times 100\% =$$

$$\frac{(1 \; amu)(12)}{(12amu)(6) + (1amu)(12) + (16amu)(6)} \times 100\% =$$

$$\frac{12 \; amu}{180 \; amu} \times 100\% = 6.67\% \; Hydrogen$$

The overall formula weight for $C_6H_{12}O_6$ is 180 amu, and the percent of Hydrogen is 6.67%.

$$\frac{(mass\ of\ Oxygen)(\#\ of\ Oxygens)}{mass\ of\ C_6H_{12}O_6} \times 100\% =$$

$$\frac{(16\ amu)(6)}{(12amu)(6) + (1amu)(12) + (16amu)(6)} \times 100\% =$$

$$\frac{96\ amu}{180\ amu} \times 100\% = 53.3\%\ Oxygen$$

The overall formula weight for $C_6H_{12}O_6$ is 180 amu, and the percent of oxygen is 53.3%.

What is the percent nitrogen, by mass, in $Ca(NO_3)_2$?

$$\frac{(mass\ of\ Nitrogen)(\#\ of\ Nitrogen)}{mass\ of\ Ca(NO_3)_2} \times 100\% =$$

$$\frac{(14\ amu)(2)}{164.1\ amu} \times 100\% =$$

$$\frac{28\ amu}{164.1\ amu} \times 100\% = 17.1\%\ Nitrogen$$

What is the percent aluminum, by mass, in $Al(OH)_3$?

$$\frac{(mass\ of\ Aluminum)(\#\ of\ Aluminum)}{mass\ of\ Al(OH)_3} \times 100\% =$$

$$\frac{(27\ amu)(1)}{78\ amu} \times 100\% = 34.6\%\ Aluminum$$

8.2

A 0.020 M solution of niacin has a pH of 2.76. Calculate the percent ionization of the niacin.

Recall that $[H^+] = 10^{-pH}$, *so* $[H^+] = 10^{-2.76} = 1.737 \times 10^{-3}$ M

$$Percent\ ionization = \frac{[H^+]\ equilibrium}{[HA]initial} \times 100\%$$

$$= \frac{1.737\ x\ 10^{-3}M}{0.020\ M} \times 100\% = 8.7\%$$

A 0.010M solution of H_3PO_4 has contains 3.7 x 10⁻³ M H⁺(aq). Calculate the percentage of acid that has ionized.

$$Percent\ ionization = \frac{[H^+]\ equilibrium}{[HA]initial} \times 100\%$$

$$= \frac{3.7\ x\ 10^{-3}M}{0.010\ M} \times 100\% = 37\%$$

8.3

$CaCl_2$ is used to melt ice on roads. A 25.6g sample was evaporated leaving a 4.36g residue. What is the (m/m)% of the sample assuming the residue contained only $CaCl_2$?

$$Mass\ percent = \frac{mass\ of\ solute}{mass\ of\ solute + mass\ of\ solvent} \times 100$$

$$Mass\ percent = \frac{4.36g}{25.6g} \times 100 = 17.0\%$$

What volume of a glucose solution would contain 0.185 g given the solution is 5.0% (m/v)?

$$mass\ by\ volume\ percent = \frac{mass\ of\ solute}{volume\ of\ solution} \times 100$$

$$\frac{mass\ by\ volume\ percent}{100} = \frac{mass\ of\ solute}{volume\ of\ solution}$$

$$\frac{mass\ by\ volume\ percent}{100} \times volume\ of\ solution = mass\ of\ solute$$

$$volume\ of\ solution = \frac{mass\ of\ solute}{mass\ by\ volume\ percent} \times 100$$

$$volume\ of\ solution = \frac{0.185}{5.0} \times 100 = 3.7\ mL$$

Commercially available rubbing alcohol contains 70% iso-propyl alcohol by volume. How many mL of isopropyl are in a 500.0 mL container?

$$volume\ by\ volume\ percent = \frac{volume\ of\ solute}{volume\ of\ solution} \times 100$$

$$\frac{volume\ by\ volume\ percent}{100} \times volume\ of\ solution$$
$$= volume\ of\ solute$$

$$volume\ of\ solute = \frac{70}{100} \times 500mL = 350.0\ mL$$

8.4

What is the percent yield of a reaction where 230 g of phosphorus trichloride reacts with excess water to form 152 g HCl and aqueous phosphorous acid (H_3PO_3)?

First we need to write the balance equation for this reaction:

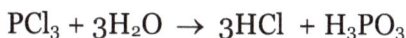

$$PCl_3 + 3H_2O \rightarrow 3HCl + H_3PO_3$$

With the amount of PCl_3 given, we can calculate the theoretical yield based on this amount for the HCl and then calculate the percent yield from there.

Game plan: g PCl_3 → mol PCl_3 → mol HCl → g HCl

$$230\ g\ PCl_3 \times \frac{1\ mol\ PCl_3}{137.3\ g\ PCl_3} \times \frac{3\ mol\ HCl}{1\ mol\ PCl_3} \times \frac{36.5\ g\ HCl}{1\ mol\ HCl} = 183\ g\ HCl$$

The theoretical yield for this reaction is 183 g, which can be plugged into the percent yield equation.

$$Percent\ Yield = \frac{actual\ yield}{theorectical\ yield} \times 100\% = \frac{152g}{183g} \times 100\% = 83\%$$

The combustion of butane, C_4H_{10}, a component of natural gas, proceeds as follows:

$$2C_4H_{10} + 13O_2 \rightarrow 8CO_2 + 10H_2O + heat$$

If 10 g of butane produce 13.3 g of CO_2, what is the percent yield of CO_2?

First we need to determine the theoretical yield, then plug it into the percent yield equation.

Game plan: g butane → mol butane → mol CO_2 → g CO_2

$$10 \; g \; C_4H_{10} \times \frac{1 \; mol \; C_4H_{10}}{58g \; C_4H_{10}} \times \frac{8 \; mol \; CO_2}{2 \; mol \; C_4H_{10}} \times \frac{44 \; g \; CO_2}{1 \; mol \; CO_2} = 30.3 \; g \; CO_2$$

$$Percent \; Yield = \frac{actual \; yield}{theorectical \; yield} \times 100\% = \frac{13.3 \; g}{30.3 \; g} \times 100\% = 43.9\%$$

When 56.6 g sulfur and 30.5 g oxygen gas undergo a reaction that has a 90.0% yield, what mass of sulfur dioxide forms?

Our first step for determining this question is to write a balance equation for the reaction.

$$S + O_2 \rightarrow SO_2$$

Since we are given quantities for each of the reactants, we need to figure out what is the limiting reactant and determine the theoretical yield based on that.

$$56.6 \; g \; S \times \frac{1 \; mol \; S}{32g \; S} \times \frac{1 \; mol \; SO_2}{1 \; mol \; S} \times \frac{64g \; SO_2}{1 \; mol \; SO_2} = 177.2 \; g \; SO_2$$

$$30.5 \; g \; O_2 \times \frac{1 \; mol \; O_2}{32g \; O_2} \times \frac{1 \; mol \; SO_2}{1 \; mol \; O_2} \times \frac{64g \; SO_2}{1 \; mol \; SO_2} = 61 \; g \; SO_2$$

Oxygen is the limiting reactant so the theoretical yield that we will use is 61 g SO_2. The question is not asking us for the percent yield, it is asking us for the actual yield so to solve this problem we will need to rearrange the equation.

$$Percent\ Yield = \frac{actual\ yield}{theorectical\ yield} \times 100\%$$

$$\frac{Percent\ Yield}{100} \times theorectical\ yield = actual\ yield$$

Now we plug in the values for the problem and get:

$$actual\ yield = \frac{90}{100} \times 61\ g = 54.9\ g\ SO_2$$

Additional Crystal Structure Problems (9.1)

Iridium crystallizes in a face-centered cubic unit cell that has an edge length of 3.833Å. What is the density of iridium metal and what is the atomic radius of an iridium atom?

$$Density = \frac{mass}{volume}$$

For the mass aspect, we know that a face centered cubit unit cell contains 4 atoms and each atom of Ir is 192.22 amu. We will need to convert this to grams.

$$mass = 4 \times 192.22\ amu \times \frac{1\ g}{6.02 \times 10^{23}\ amu} = 1.2772 \times 10^{-21} g$$

For the volume aspect, we know that the volume is the edge length cubed. The edge length is given in Å, but we need the volume measurement to be in terms of cm³, so we will need to convert Å to cm.

$$3.833\ \text{Å} \times \frac{1 \times 10^{-10}\ m}{1\ \text{Å}} \times \frac{100\ cm}{1\ m} = 3.833 \times 10^{-8} cm$$

With the edge length in terms of cm, we can now calculate the volume.

$$volume = (3.833 \times 10^{-8} cm)^3 = 5.631 \times 10^{-23} cm^3$$

To solve the density, we need to take the mass and divide it by the volume.

$$Density = \frac{mass}{volume} = \frac{1.2772 \times 10^{-21}g}{5.631 \times 10^{-23}cm^3} = 22.68 \ \frac{g}{cm^3}$$

For the second part of this problem, we are asked to determine the radius of Iridium. Since we are given the packing and the edge length we can determine the radius.

FCC packing has an edge length of $\frac{4r}{\sqrt{2}}$ and is equal to 3.833 Å. We will need to rearrange to solve for r.

$$\frac{4r}{\sqrt{2}} = 3.833 \ \text{Å}$$

$$4r = 3.833 \ \text{Å} \times \sqrt{2}$$

$$r = \frac{3.833 \ \text{Å} \times \sqrt{2}}{4}$$

$$r = 1.355 \ \text{Å}$$

Calcium crystallizes with a body-centered cubic structure. How many Ca atoms are contained in each unit cell? If the radius of calcium is 1.97Å, what is the length of the unit cell? What is the density of Ca metal?

Body-centered cubic structures: 1/8 atom x 8 corners plus 1 in the center.

$$\frac{1}{8} \times 8 + 1 = 2 \ atoms \ per \ unit \ cell$$

Based on Table 4, body-centered cubic unit cells have an edge length of $\frac{4r}{\sqrt{3}}$ so we will need to plug in the radius value into this equation to get the length of the unit cell.

$$edge \ length = \frac{4 \times 1.97\text{Å}}{\sqrt{3}}$$

$$edge \ length = 4.55\text{Å}$$

For the density:

$$mass = 2 \times 40.08 \; amu \; \times \frac{1 \, g}{6.02 \, \times \, 10^{-23} amu} = 1.3315 \times 10^{-22} \, g$$

$$volume = (edge \; length)^3 = \left(4.55\text{Å} \; \times \frac{1 \times 10^{-10} m}{1 \, \text{Å}} \times \frac{100 \; cm}{1 \, m}\right)^3$$

$$= 9.4196 \times 10^{-23} cm^3$$

$$density = \frac{1.3315 \, \times \, 10^{-22} g}{9.4196 \times 10^{-23} cm^3} = 1.41 \frac{g}{cm^3}$$